• ITN BOOK OF FIRSTS •

By Melvin Harris

MICHAEL O'MARA BOOKS LIMITED

TO MY FIRST LADY – FIRST IN ALL THINGS!

First published in Great Britain in 1994
by Michael O'Mara Books Limited
9 Lion Yard, Tremadoc Road
London SW4 7NQ

A CIP catalogue record for this book is available from the British Library

ISBN 1-85479-199-0 (hardback)
ISBN 1-85479-737-9 (paperback)

Designed and typeset by
Florencetype Limited, Kewstoke, Avon

Printed and bound in England by
Clays Limited, St Ives plc

10 9 8 7 6 5 4 3 2 1

• CONTENTS •

PICTURE ACKNOWLEDGMENTS

All pictures kindly supplied by the Maureen Gavin Picture Library

• PREFACE •

Writing a book of FIRSTS is a perilous venture. Lurking in some dusty archive in some battered notebook, in some rusting tin trunk, may well be an item which rewrites history. But never mind, as long as the record is made without dogmatism or chauvinism – every country likes to ascribe important inventions to one of its own sons or daughters – it cannot fail to be instructive and even entertaining: given, of course, that the writer is not content with a dry list of facts.

When reading this book be warned that not every reference book can be relied on. If you spot an entry that seems to contradict one of mine, please double-check. Perhaps even check once more, for writers tend to be lazy and pick up a great deal of their 'facts' second or third hand. That is why at least five reference books tell us that bassoon-maker Heckel also invented a woodwind called the Heckelclarind. You would have a hard job finding one of these anywhere, for what Heckel invented was the Heckelclarina, but a misprint in one book was slavishly copied and recopied and so a mythical instrument was born.

One other note of caution: all patent specifications should be examined with great care, since they often incorporate details added after the original date of application.

Finally, note that in my text I herald the first appearance of an invention or trade-name by placing it in italics – after that I treat it as a grown-up and leave it to fend for itself!

• GLORIOUS STEAM •

Steam was first harnessed in Egypt as long ago as the year 150 BC. Egypt, then a Roman province, had Alexandria as its capital, and one of the Alexandrian citizens was the Inventor Ctesibus. He was the father of many ingenious pneumatic and hydraulic machines. This man probably invented the first steam turbine, the *Aeolipile*, a neat engine illustrated in Hero's *Pneumatics*. Steam, driven into a copper globe, squirted out of two small tubes fixed opposite to each other and bent at right angles. As the bends faced in opposite directions, and as the globe was free to revolve on its pivots, the escaping steam jets drove it round at an impressively high speed. It demonstrated the awesome power of steam, but we have no direct proof that it was ever used to drive any other apparatus. Perhaps it was cheaper to use manual labour and whips!

But this story doesn't end there, its climax came 2,000 years later, in 1845. In that year the Alexandrian 'Aeolipile' was first used for work beyond doubt, when Sir Arthur Cotton employed such a device to drive the fan providing air-blasts to his iron-foundry furnace. He was delighted by its basic simplicity and unmatched efficiency. It was the first of his range of drivers adapted to iron smelting.

Such a simple prime driver has very limited uses and it was certainly not considered when a mineshaft pumping engine was being sought in the seventeenth century. English miners, at that time, were increasingly menaced by underground water. As their shafts went deeper so they had great difficulty in clearing them of the vast quantities of water which seeped, trickled and even gushed in. Thomas Savery, of Devonshire, worked at a solution and on 25 July 1698 he patented the first steam-powered pumping engine. His first working model of the invention was shown to King William III in 1698 and his first production model was adapted to raising water for drinking and watering gardens. His later mine-pumping engines were the first commercially valuable steam-engines in the world. Known as the *Miner's Friend* his engine could replace the huge teams of horses normally used to raise the water in buckets.

It was a crude start but it fired others to work on improvements and in 1705 Thomas Newcomen and John Calley were able to give a detailed account of their new *Atmospheric Steam-Engine*. This used the cylinder

The 'Aeolipile' of Antiquity

and piston combination (first used by Papin in June 1690) and harnessed the driving power to an overhead beam. This beam took the piston movement at one end and the pump rod at the other. It was like a giant see-saw in its actions. It counts as the first successful piston engine of commerce, since after its first use in 1712 (at Griff), it was introduced in nearly all the large mines in Britain. By the end of the century improved Newcomen engines were pumping water from waterworks to houses and emptying dry docks, including the great dock at Kronstadt in Russia (first installed 1777); while in Holland the engines were used to drain the waterlogged lands (first used 1777). It was also the first steam-engine used to operate the blowing-machinery at a blast-furnace (Carron's of Falkirk 1765).

Newcomen's engine, or rather a *model* of his engine inspired James Watt to create his own engine. The model was given to Watt for repairs to be made on it. On testing it he found it underpowered and began to ask why? He discovered that too much heat was lost from the cylinder, and he remedied this with a condensing system that kept the cylinder as hot as possible. This separate condenser liberated the steam-engine. It was first used in Watt's model of August 1765, then used in the first full-sized engine of September 1769. But there were still many imperfections. 'Of all things in life', wrote Watt, 'there is nothing more foolish than inventing.' But at a moment of deep despair he was saved by an alliance with Matthew Boulton. Boulton had been working on his own engine and had actually shown a model to Benjamin Franklin. Yet he quickly recognized that Watt's ideas were in advance of his. A partnership was arranged and in November 1774 Watt gave the first successful trial of his Kinneil engine; the first truly modern steam-engine in all its essentials.

It has been said that from then on nearly every important invention that marked the history of steam power originated in the fertile brain of James Watt. These include cross-heads and guides, poppet valves with bevelled seatings, air-cooled condensers, throttle-valves, steam-gauges, crank-drives and the fly-ball governor. All these first originated with Watt. Apart from that he also invented the steam hammer. Yet he would have been the first to acknowledge the ingenuity of his friend and helper William Murdoch. Murdoch was first to apply a sun-and-planet drive to Watt's 'rotative engine' and in 1808 he was first with a scheme to use coal-gas for lighting streets and houses. Ten years earlier he had actually introduced the first coal-gas lighting at Boulton's works. He further invented the first steam gun (later used in the American Civil War) and the first pneumatic dispatch tubes to send packages from one place to another, using the power of compressed air. Murdoch was also the man who took Richard Trevithick under his wing when the young apprentice came to learn all about pumping engines. Having studied the stationary engine, Trevithick then turned to the idea of a locomotive engine. First it was a road vehicle that he dreamed of, then he switched his affections to a locomotive able to run on rails.

On 22 February 1804 Trevithick demonstrated the world's first steam-powered railway locomotive. He had made a full-sized loco a year earlier, but *Number Two*, on that day in February, hauled 10 tons of iron, plus seventy men and five extra wagons for a full 15km almost 9.5 miles. His speed was a contented 5mph. The only drawback at that time lay in the frailty of the rails then used. They were not robust enough to bear the weight and the wear and tear that the new 'iron horse' necessitated. This arose because railways first came into being as tracks for wagons that were either pushed by hand or drawn by horses. For this type of use wooden rails were mainly employed. Stronger, cast iron rails were first made in Sheffield in 1776, but even these were not meant to take great weights. And it was a broken rail that led to the end of the first public railroad open to paying customers. This was yet another of Trevithick's babies. In London's Torrington Square, (Euston Square), he built a circular railway track and around this track he drove an engine named *Catch Me Who Can*. The engine pulled a single open carriage full of brave travellers. Hour after hour it trundled around on its journey to nowhere, until a rail cracked and the engine was toppled. Lack of funds prevented a new start and the inventor turned to fresh challenges. It was left to Matthew Murray to build the first commercially successful railway.

In 1812 Murray fitted a locomotive with a large toothed driving wheel. This cogwheel gripped the rack of teeth fixed at the side of each rail. The toothed drive was an idea of his partner, Blenkinsop and though a dead-end in itself, it made the Middleton railway (near Leeds) a sound proposition. It pulled the coal wagons with ease and it cost less than the old-style teams of horses. But this was simply a goods service. The first passenger service had to wait for the genius of Robert Stephenson.

Stephenson began by designing and making colliery engines; his first important locomotive first took to the tracks in July 1814. This was his *Blucher*, an engine fitted with four plain, flanged wheels. Just five months earlier William Hedley's *Puffing Billy* loco had demonstrated for the first time that smooth wheels would adhere to smooth rails and provide traction. This freed designs from dependence on the slow and noisy rack-drive system. The only remaining snag still lay with fragile rails. Stephenson's engines (four in all) did sterling service as colliery work-horses, but their ambitious designer was looking much farther ahead. He concentrated on new ways of making tracks dependable and hard wearing, and in October 1820 he greeted the arrival of John Birkinshaw's patent rails with enthusiasm and relief. These rails were the first dependable wrought-iron rails of the century. They set the standards for many years to come. They came in 18-foot lengths, had a swelled upper edge and the ends were welded.

Birkinshaw's patent rails were used on the first public steam railway, the Stockton and Darlington Railway, which first opened on 27 September 1825. The first locomotive to run that day was Stephenson's *Locomotion*, a new design of his using lap-and-lead and outside coupling rods for the first time. The railway itself was started as a goods service only. Passengers had to wait until 15 September 1830 before they were made welcome on a steam line. On that day the Liverpool and Manchester Railway was formally opened. It was a day of mixed rejoicing and tragedy, for everything went perfectly until Home Secretary Huskisson was knocked down by one of the engines and died of his injuries. This was the first fatality on steam railways.

Drawing the first carriages on that day was Stephenson's *Rocket*, the winner of the national competition fought out at Rainhill in 1829. The Rainhill trials had put rival locomotives to the test and the *Rocket* showed itself to be reliable, efficient and way ahead of any other loco that existed. These trials of 4–6 October 1829 were the first to prove the value of steam for a passenger service. There had been a passenger rail service before this in the shape of the Swansea and Mumbles Railway. This service opened up on 25 March 1807, which makes it the first of its kind, but its carriages were *horse-drawn* until 1877, so the Liverpool and Manchester is the first modern railway, without any dispute. And it was the first to run a passenger service to a time-table. Its double tracks were the first as well.

In the United States the first railroad had been constructed in 1795 on Beacon Hill, Boston, Massachusetts. Like the first European roads it was never meant for steam. Steam only came into the picture in 1823, when a charter was granted for a steam-powered service between Philadelphia and Columbia, Pennsylvania. At that time there were no native built locos in the States. The first was not made until 1825. This first, by John Stevens of Hoboken, worked on a rack system and had four flat-rimmed wheels guided by upright rollers. It had no future but it focused thought on the possibilities of steam. Thought became action when the Delaware and Hudson Canal Company imported a locomotive from Foster, Rastrick and Company of Stourbridge, England. This engine, the *Stourbridge Lion* was first tried on the American lines on 7 August 1829. But the Americans had the same rail problems as their British cousins, so the *Lion* made nothing more than trial trips. A much lighter, American-made locomotive became the first to run easily over the primitive tracks. Called *Tom Thumb*, it was a quaint looking affair with an upright boiler mounted on a flat cart. It was the work of Peter Cooper, who later founded the famous Cooper Union in New York. It first ran on 25 August 1830, and despite its size, it developed three times as much power as the *Rocket*, and covered 13 miles in less than an hour. It demonstrated that the curves and heavy gradients of the Baltimore and Ohio system could be taken safely at high speeds. Within a year the Baltimore–Ohio became the first railroad in the USA to carry troops (30 June 1831) and on 6 June 1833 it became the first to carry a President of the USA (Andrew Jackson). It followed this by being the first to run a train into the capital city of Washington, on 24 August 1835.

Though *Tom Thumb* had demonstrated useful principles it was little more than an experimental loco. The first successful US loco to reach reliable commercial standards was the *Best Friend of Charleston* which went into passenger service with the South Carolina Railroad on 14 December 1830. Full service then began on Christmas Day. Sadly the

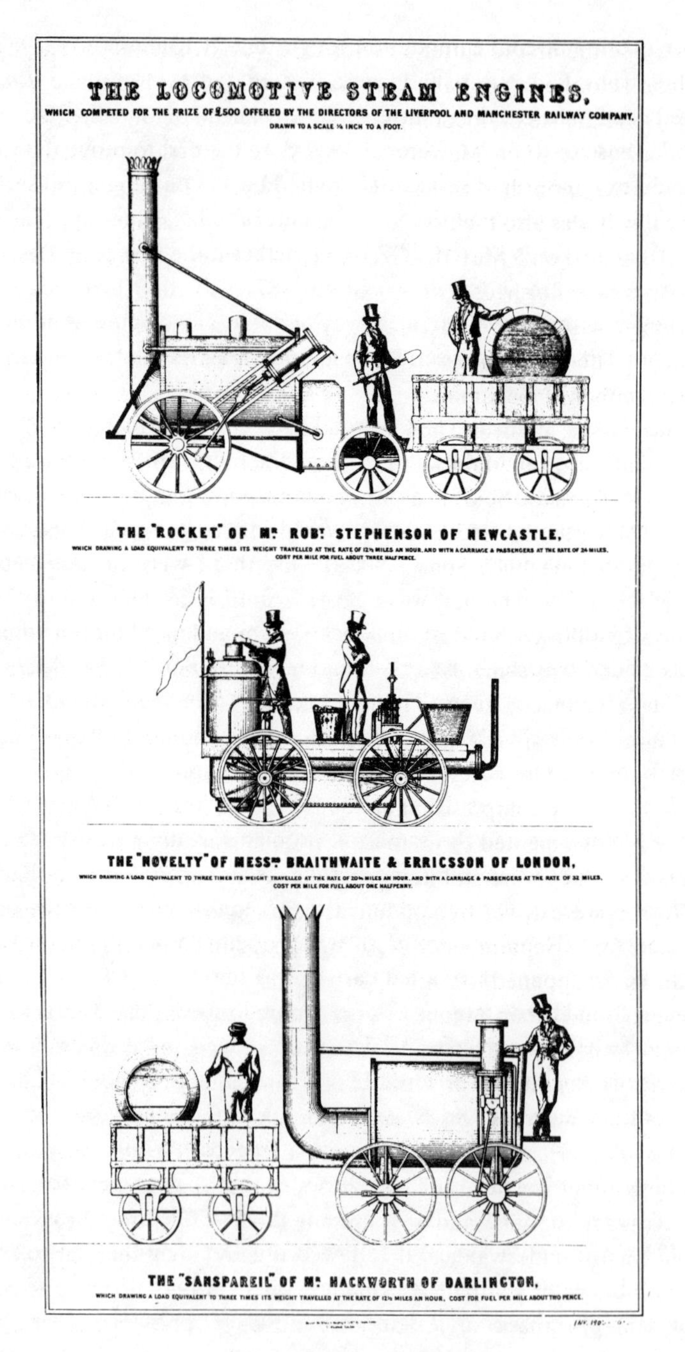

The crucial Rainhill Trials

loco became the first to suffer a boiler explosion, when on 17 June 1831, a fireman held the safety valve down, and suffered for his folly.

In Europe the steam-locomotive made its first showing in France on 7 November 1829, when Marc Seguin tried out his multi-tubular boiler engine on the unfinished St-Etienne–Lyons tracks. In Belgium their first steam railway was also the first in Europe and the first state railway in the world. It opened on 5 May 1835 when the 14.5-mile stretch from Brussels to Malines was finished. Two of its locomotives were built by Robert Stephenson, while a third, called the *Olifant* was the first to be built on the Continent. The first German steam railway company, the Ludwigsbahn, chose a Stephenson locomotive *Der Adler* to haul its trains along its lines from Nuremberg to Furth. The first journey was on 7 December 1835. And the Russians also relied on a Stephenson engine to work their first lines (Pavlosk to Tzarskoe Selo) in December 1836. Other first openings were: Holland, 24 September 1839; Italy, 4 October 1839; Hungary 15 July 1846; Denmark, 26 June 1847; Spain, 28 October 1848; Norway, 1 September 1854; Portugal 28 October 1856; and Sweden, 1 December 1856. In Switzerland railway buildings faced special problems in the Alpine regions. When tunnelling at St Gotthard began some 400 cubic metres of debris were produced every day and dumped into trucks placed on the rails. Steam locomotives could not be sent into the tunnels because of the lack of ventilation, so a new type of locomotive was created. These were the first driven by compressed air which was stored in the former boiler space and supplemented by extra air carried in a cylinder bolted to a tender. Even so, they were still ultimately dependent on steam, since the air compressors were driven by stationary steam-engines. Work first began on this tunnel on 13 September 1872. Its two bores first met on 13 September 1880 and it first opened for traffic on 1 January 1882.

No tunnel in Britain ever presented such problems. The first used for passenger work was a paltry 838 yards long, compared with the 9 miles 662 yards of the tunnel at St Gotthard. Britain's first was opened on 4 May 1830 at Tyler Hill on the Canterbury and Whitstable line. But the first underwater railway tunnel was a very different proposition. Stretching under the Severn between South Wales and England, it took fourteen years to make and at 4 miles and 628 yards long, became the world's longest underwater tunnel. It first opened on 1 September 1886.

The problems of ventilation in tunnels gave an impetus to those who believed in pneumatic or atmospheric railways. The first ideas along these lines date back to 1824, when John Vallance patented his atmospheric system (No. 4,905). The great Brunel, usually a shrewd judge of viable inventions, fell in love with the atmospheric railway but came down to earth when his first trains proved unreliable. His venture, first tested on the South Devon Railway became known as the *Atmospheric Caper*. Like all such railways it could only work if the driving tubes, containing either a vacuum or compressed air, were leak-proof. His were not. Rats made a feast of his greased leather seals and the weather joined in to warp them out of fit. A different fate met the pneumatic dispatch system organized by the Electrical and International Telegraph Company. Their first dispatches were sped through a tube from Lothbury to the Stock Exchange in 1854. The motive power used was compressed air and the system worked without a hitch. Its first tubes, though, were just an inch in diameter! Still they did lead to bigger things. First the Post Office took up the idea and opened its own pneumatic service in February 1870. It then took the next step and had built a tube network large enough to take small carriages. These were used to convey mailbags and parcels from one large sorting office to another, pneumatically. The first part of this system opened in February 1863, using 2-foot-6-inch-diameter tubes. A change was then made to larger tubes of 4 foot 6 inches, and this improved service made its first run in November 1865. But there were still problems with leakage of air and the scheme was abandoned. Its partial success later led to the present-day Post Office underground system which was first authorized in 1913, held up by the First World War and only saw its first service in December 1927. Wind power was now replaced by electricity.

In the USA pneumatic power was not reserved for mailbags. On 26 February 1870 the first underground railway in America opened up in New York City. Its first tunnel ran under Broadway and connected Warren Street with Murray Street. Its passenger car was driven through the tunnel smoothly and safely by a gigantic blower engine. Behind the blower, though, was a steam-engine of 100 horsepower. For in the end all pneumatics depend on the power of steam. In fact behind every prime mover, whether oil, gas or electrically driven, there lurks the power station. And no matter what fuel a power station consumes, it is invariably used to drive steam-engines or steam-turbines.

It was, of course, as a means of navigation that steam first became a conqueror of distance. The first written suggestion of the use of steam to drive a ship dates from 1690, when Papin's paper *Acta Eruditorum* referred to '... oars fixed to an axis ... most conveniently made to revolve by our tubes. It would only be necessary to furnish the piston-rod with teeth, which might act on a toothed wheel ... which would communicate a rotary motion ...'

Papin's 'tubes' were steam-pumping engines, which worked well enough, but his idea of oars was soon overtaken by Thomas Savery's patented paddle-wheel ship of 1696. In 1736 Jonathan Hulls patented his own paddle-wheel steamboat meant for towing work. But this never went further than forming the subject of an amusing illustration in Hulls's first published pamphlet in 1737. It was not until 1774 that a first attempt was made to couple the paddle-wheel with the steam-engine. This took place on the Seine and the people behind the vessel were Comte d'Auxiron and M. Perrier. Alas, their vessel was grossly underpowered so their experiment was valiant but premature. A further attempt at Lyons on the 15 July 1783 is reported to have been a triumph, but the Academy in Paris, miffed by a mere provincial exploit, refused to rejoice and the discouraged navigators gave up striving. The first real success with a steamboat was gained by William Symington on 14 October 1788. He co-operated with James Taylor and fitted a small steam-engine to a twin-hulled boat made along the lines set out by Patrick Miller in 1787 (today we would call such a ship a 'catamaran'). This first steamer, 25-feet long and 7-feet broad, was tried out on the lake at Dalswinton and delighted everyone, yet it only received brief notices in two newspapers and one magazine (*Dumfries Newspaper*, *Edinburgh Advertiser*' and *Scots Magazine*). The *Annual Register* for 1788 does not even bother to mention this first successful steam navigation. Fortunately its engine survived many hazards, including a projected visit to a scrapyard, and can still be seen at London's Science Museum.

In 1802 Symington built the first commercial steamboat intended for canal work. It was commissioned by Lord Dundas and named *Charlotte Dundas*, after his daughter. It took over the duties of the horses used to tow the barges, but had a short career since its wash damaged the banks of the canal. It was the first boat to use a cranked axle to give rotary motion to drive the paddle-wheel; all other attempts had involved the use of levers or beams.

In the United States as early as 1784 James Rumsey had been pressing for steam navigation to be taken seriously. He tried to animate George Washington himself, but met with apathy. In September 1785, John Fitch presented a working model of his proposed steamship to the American Philosophical Society at Philadelphia. His first proposal was followed, in March 1786 by the granting of the first exclusive patent to steam-navigate the waters of Philadelphia. He was granted a monopoly for fourteen years. A similar monopoly was granted New York on 19 March 1787; by Delaware in February 1787, and by Pennsylvania on 28 March 1787.

With those four first licences to his credit, Fitch then put his first boat on trial in May 1787, found it inadequate and modified it to the point where he could then offer it for testing before a most distinguished group of observers, the delegates to the Convention then drafting the Federal Constitution. They met at Philadelphia on 22 August 1787 and Fitch left with many letters vouching for his success. This first top-level endorsement led to a further patent from Virginia (7 November 1787) and the construction of his first large ship (60-feet long) in July 1788. This new craft had a boiler failure on its first trip, but it did make a full 20 miles under its own power. It returned by floating with the tide. On 26 July 1790 more improvements led to the first boat placed in service as a passenger-boat. This first shipping line ran between Philadelphia and Trenton, taking in Burlington, Bristol and Bordentown, with occasional trips to Wilmington and Chester. The first fares were 2s 6d to Burlington and Bristol, double that to Trenton, and 3s 9d to Bordentown.

The Fitch service was abandoned after a couple of years and no attempt was made to replace it until Robert Fulton came along with his paddle-steamer *Clermont*. Launched in spring of 1807 *Clermont* made the first steam-vessel journey of any considerable length in August 1807. She ran the 150 miles distance from New York to Albany in thirty-two hours and then made the return journey in thirty hours. This was the start of the first commercial service that ran on a regular basis. As this new craft passed she overwhelmed the onlookers with her presence. She was unfamiliar, exciting; one man described her as 'A monster moving on the waters, defying wind and tide, and breathing flames and smoke.' The dry pine wood she burned helped create the fire-dragon illusion. It was said that some seamen used to hide below deck at the approach of the ship. The first fares on the *Clermont* ranged from $3 to $7, and the first booking agent for the trips was William Vandervoort of No. 48 Courtland Street, corner of Greenwich Street, Pauler's Hook Ferry.

The *Clermont*, first steam-vessel to make a voyage of considerable length

In 1812 Fulton diversified by introducing the first steam ferry-boat; it worked between New York and Jersey City and used a twin-hull construction. It could transport 8 carriages, 30 horses and still have room for up to 400 passengers. It was the first vessel with such a capacity. In the same year the War of 1812 led Fulton to design the first steam war vessel. His plans were first approved by Congress in March 1814 and she was launched on 29 October 1814. In July 1815 she made her first trial-trip. In September she took on board her battery of guns and provisions and steamed to Sandy Hook and back, some 53 miles at the rate of 5.5 miles an hour. On board she carried thirty 32-pounder guns intended to fire red-hot shot and a collection of large pumps meant to douse an enemy's decks and turn his powder into a grey, worthless sludge. This first fighting steamer bore the name *Fulton the First*.

In his designs Fulton stayed loyal to the paddle-wheel, but in 1804, one of his rivals, Colonel John Stevens of Hoboken, completed a steamboat driven by a *screw-propeller*. This first screw-driven boat was also the first to use a viable sectional boiler. This boiler used 100 tubes, 2 inches in diameter, to carry the water over the furnace flames. This made for efficiency and allowed a head of steam to be raised at a fast pace.

Colonel Stevens followed his first boat with the first twin-screw vessel in 1805. He then devised the first iron-clad ship which doubled as the first revolving gun turret! It was a saucer-shaped vessel with its screw-propellers so arranged that the whole ship could be rotated around on its axis. Its guns were set around the deck like the spokes of a wheel. When anchored at a harbour mouth it could fire each gun in turn as the ship revolved. This was the first idea of the 'Monitor' principle, which, after a while, became important. But when it was first proposed the plans were shelved. They were then taken seriously by the Russian Government who introduced the ship into their navy under the name *Popoffka*.

The Colonel's son, Robert, joined in his father's work as a boy and in 1818 he became the first to successfully use anthracite coal for steam production. Before then it was regarded as an almost unmanageable fuel, but in the special furnaces of his steamboat *Passaic* he showed that it could give out great heat with little waste. Anthracite became known as *the* steam coal and came into international use after this date. Yet another first for the Stevens came in June 1809 when their ship *Phoenix* made the first seagoing journey by steam. Setting off on 10 June, *Phoenix* left New York, headed out to sea and arrived at Philadelphia thirteen days later.

In Europe practical steam navigation began in Scotland when Henry Bell, of Helensburgh constructed his steamer *Comet* in 1812. On 5 August 1812 he opened the first passenger service in Britain with a circular to the public issued from his offices at Helensburgh Baths. The service operated between Glasgow, Greenock and Helensburgh; those first fares were 'four shillings for the best cabin, and three shillings for the second, but beyond these rates nothing is to be allowed to servants or any other person employed about the vessel'. Nothing was said about compensation for hard labour, though a contemporary account tells us that 'It was no uncommon occurrence for the passengers, when the little steamer was getting exhausted, to take to turning the fly-wheel to assist her.' And it was a very little steamer, just 40 feet long with an engine of just about 4hp, but for all that, it was the first. Its all-wood construction was a legacy from the past, and it was not until 30 April 1822 that the world's first iron ship took to the water. This 116-ton ship, the *Aaron Manby* made her first trips on the Thames, then made her first maiden voyage over to Paris on 10 June 1822.

Just three years earlier a much smaller wooden steamship made the first crossing of the Atlantic. The American steamer *Savannah* left port at Savannah, Georgia, on 24 May 1819 and docked at Liverpool twenty-seven days later. But this 50-ton vessel only spent part of her time under steam, and the first crossing using steam all the way was not made until 17 August – 4 September 1838. This journey, from Pictou, Nova Scotia to Cowes on the Isle of Wight was also the first to cater for passengers. The brave fare-payers were eight in number; the first cargo included a collection of stuffed birds and a harp, and the ship was the Canadian *Royal William* built in Quebec in 1831. After that, the race was on to win the Atlantic for steam power. In 1838 this reached the stage of an actual race between two British steamers. On 4 April 1838 the London–Cork ferry boat, *Sirius*, a 700-ton worthy, set out for New York with her forty passengers. She arrived in New York on 23 April 1838 just hours before her rival the *Great Western*. But although this made *Sirius* the first from Britain, the *Great Western* had in fact, made the journey at a faster rate, taking three days less to cross. And *her* voyage, (her first ever) marks the start of the first regular transatlantic service for passengers. Her sister ship, the *Great Britain*, then became the first to cross the Atlantic using screw propulsion. As well as that Brunel's *Great Britain* was the first iron steamship to make the crossing. This double first began at Liverpool on 26 July 1845 and concluded at New York over fourteen days later. At that time she carried 600 tons of cargo and sixty passengers. At a much later time she became the first Atlantic ship of that period to be preserved. She lies now in the very place that saw her off on her first dip into water. There, at Bristol, she rests as the centre-piece of a maritime history display.

The transatlantic race of 1838 seemed to prove that both paddle-wheels and screw-propellers were equally suited for sea-going. But the disadvantages of the paddle-wheel soon became clear. Warships in particular needed a clear and free sweep for their guns. The screw provided this. And the screw allowed the machinery to be placed well below the waterline, leaving the upper decks free for gun laying. The screw-propeller itself is usually held to have been first created by the Swedish engineer John Ericsson, in 1837. But in May 1804 J. Stevens (US) employed a type of screw, and Trivethick patented another form of propeller in 1815. The French claim it for Frederick Sauvage of Boulogne and I am quite sure that the Russians have a name somewhere. The tangled truth is best dealt with by acknowledging that the screw-propeller was in the minds of a number of fine thinkers at roughly the same time, these include Thomas Pettit Smith and Robert Kinder. But the first man to show the screw as an impressive advance was Captain Ericsson. Ericsson gave such a convincing demonstration of propeller virtues that he scooped up an order for two boats to be sent to the United States. That first order was placed on his own account by Captain Robert F. Stockton of the US Navy in 1839. And the first ship to demonstrate Ericsson's screw was his *Francis B. Ogden*, named after the the US consul at Liverpool, who helped fund the construction. Ogden's help and the meeting with Stockton led Ericsson to the United States, where his machinery was installed in the USS *Princeton*, the first American warship to be screw-driven. From then on, the screw-propeller was introduced with great rapidity, both in the naval and civilian shipyards. It changed the whole future of travel by sea, yet the first Admiralty 'scientific' reports had cautioned against its use, on the specious grounds that a ship so propelled would be unsteerable. Gloom about the use of propellers became so infectious at one time, that actor Henry Morford even composed the first poem about them. It was a farewell dirge:

Though Billy Florence o'er the sea
Ran off last week, per mail Cunarder
A different fate remained for me—
A narrower and harder.
Gotham's Wise Men obtained a bowl
When they went gayly seaward rowing;
I, dense of body as of soul,
Astride a screw am going.

Morford later apologized: 'When I wrote that farewell poem, to my fancy the screw steamer was a mere apology as compared to the paddle . . . Billy Florence had said to me . . . "I wouldn't cross in a screw if I had to stay at home the balance of my life!" . . . Now for what experience has taught me . . . the screw steamship, full of power and full-rigged, is the perfection of a sailing ship with propelling power added . . .' Even so it took until 1853 before Samuel Cunard even admitted that screw propulsion was inevitable. And even then he waited apprehensively before bowing to the inevitable until Death decided to resolve the problem for him. Because of his indecision the first Cunarder to be screw-driven was not seen until 1884, when the *Umbria* was launched. But by then Cunard was so out of touch with the developing technology that they ordered single-screw driving. Hard lessons were dealt out by the tough Atlantic and the line soon realized that floating palaces needed twin screw-drives.

By the end of the century twin screws and a row of funnels became the hallmarks of the great liners to come. Then some odd wires began to appear on the masts of one US ship, the *St Paul*. They were there to oblige a Signor Marconi who was wondering if an invention of his could be of use to the shipping line and to ships everywhere. By the time the ship reached Southampton every passenger had bought a copy of *The Transatlantic Times*. Printed on ship and priced at one dollar, it was the first ocean newspaper to carry news obtained by wireless. And it was naturally, printed with the aid of glorious steam.

• THE OPEN ROAD •

The quest for the Horseless Carriage set many an inventor's mind in a whirl. A sail hoisted on a wheeled chassis would work at times: if the wind had power; if the ground was smooth and if the ice was thick and strong. But wind-power on land was fickle, at any time you could be left immobile and frustrated while the horse cantered by. A dubious claim for the first horseless carriage involves a vehicle demonstrated by Christopher Holtum in 1711. It was shown under the Piazzas in Covent Garden and described as able to '. . . go for five or six miles an hour'. How it worked was never explained, so it has to be dropped from the reckoning.

The first horseless vehicle seen to work and trundle along roads was a steam-driven three-wheeled affair made by Frenchman Nicholas Joseph Cugnot in 1770. He meant it to be a tractor for use in hauling guns and ammunition for the artillery. But it was underpowered, overweight and deadly to steer, so it ended up as a museum piece and can still be admired today. It was left to Cornishman Richard Trevithick to make the first successful steam carriage in 1801. He started in 1796 by making his first reduced-sized models. They worked well and prompted him to go further with a man-carrying model. This made its first public run on Christmas Eve and carried several persons. It took its load of at least one ton and a half up a steep hill at some 4mph. On the flat it ran at 8–9mph. Four days later the vehicle went up in flames while Trevithick dined and drank. No matter, the point had been proved and his improved steam coach became the next venture. This first ran in 1803 and reached speeds of 9mph with ease, and it made a memorable 10-mile circular journey through the streets of London, yet it never undertook any longer distances.

The first long-distance, steam road-vehicle was invented by Goldsworthy Gurney in 1829. It made the first long-distance run by a self-propelled vehicle when it travelled from London to Bath. Elsewhere the first self-propelled carriage had emerged in the USA in 1805. But this *Amphibious Digger*, made by Oliver Evans of Maryland was ill-suited to long distances and even in short runs was far too clumsy to catch on.

While most designers toiled over steam boilers, a few looked for a new-style prime mover. Most promising of such new ideas involved the use of a gas engine. In 1807 Swiss inventor Isaac de Rivaz made and patented a

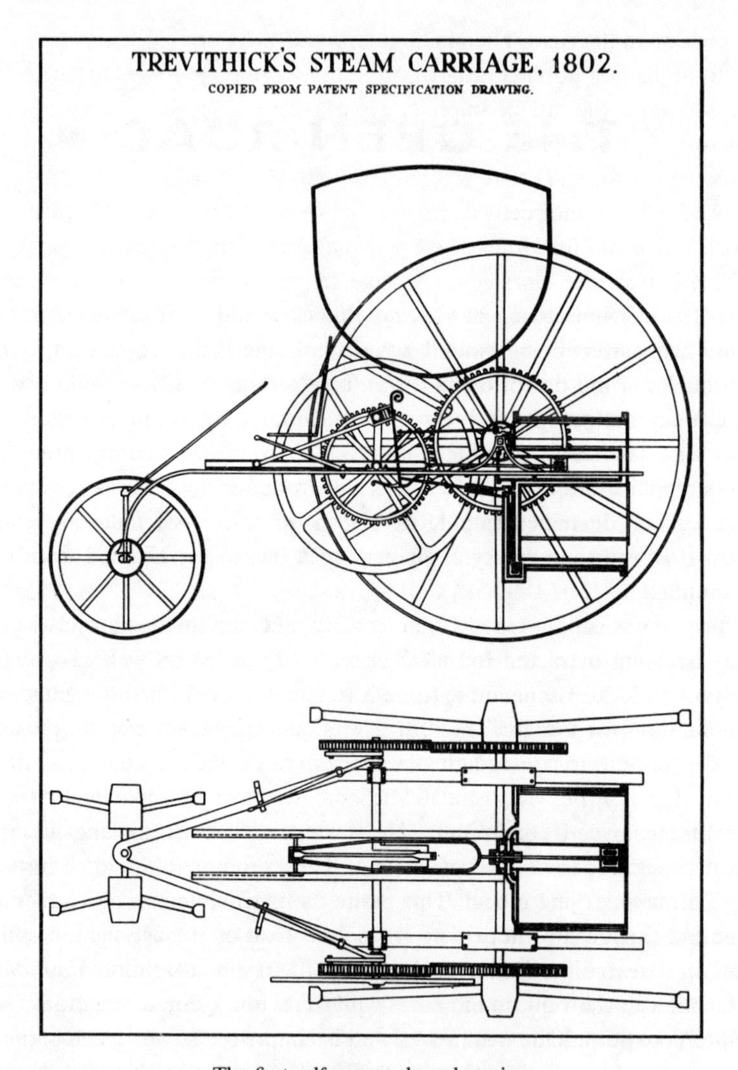

The first self-powered road carriage

powered trolley using a gas engine. It was too inefficient to develop into something of great use, but it did make a point in favour of the internal combustion engine. It led to a development of the first practicable gas engine in 1853. This was patented by Eugenio Barsanti and Felice Matteucci, but they never took their designs through to the working stage. It was the Belgian engineer J. Etienne Lenoir who developed the first efficient gas engine, though it only delivered a feeble 2hp; that was in 1860. But feeble as it was this did not stop Lenoir from drawing up plans to fit the engine in a road vehicle. Finally in May 1862 he produced the world's first motor car. Being cautious, he tested it out in the factory before taking it out on to the road. Then in September 1863 he drove the first motor car out on its first public journey on the open road. He took the car for a 6-mile spin at the rate of 4mph. A trip that took all of three hours to make. Surprisingly, despite a claim that Czar Alexander II bought this car, (the first royal driver?) that was the end of Lenoir's direct contribution to motoring. But indirectly he influenced the work of Nikolaus August Otto, creator of the first four-stroke internal combustion engine, patented 17 May 1876.

Otto's engine used half the amount of fuel consumed by Lenoir's, but was crippled by being thought of as a stationary unit. Then there arose a problem at the patent office. From out of dusty obscurity someone dug up an earlier patent taken out by Frenchman Alphonse Beau de Rochas in 1862. This showed that Rochas was the first man to describe the important four-stroke cycle. After two years of court battles, Otto's master patent was deemed invalid and other inventors were now free to use the basic ideas with impunity. Four-stroke development was then refined and modified by Karl Benz of Mannheim. Benz moved away from dependence on gas supplies; a portable engine needed a fuel similar to the new fluids being extracted from the crude oil being raised in Pennsylvania. Fluids like kerosene or gasoline. As early as 1879 he had made his first two-stroke engine; now, free from patent restraints, he used that experience to create a single-cylinder engine which he fitted into a three-wheeled vehicle. This motor car, the first successful petrol-powered example, made its first public run on 3 July 1886 at Mannheim. That first trip was made at a speed of 9mph and covered just over half a mile.

Meanwhile, just 60 miles away in Cannstadt, Gottlieb Daimler was adding the finishing flourishes to his four-wheeled petrol-engined vehicle, which became the first of its type when it ran in August 1886. By 1889 Daimler had produced the first car using a two-cylinder engine built on the V pattern. One year earlier Benz had sold his first commercially conceived car to a French customer, but unlike Daimler he still clung to his original three-wheel pattern. It was not until 1893 that Benz produced his first four-wheel car in direct competition with Daimler.

It was a Benz model that was first used as the basis for the first motor car made in England. The maker was Walter Arnold of Peckham, South London, who had imported the first Benz car in November 1894. (Another Benz original made the first British public highway run in November 1894.) Having secured a licence from Benz, Arnold then modified the Benz engine to give greater power and used this in their first Arnold motor car, which made a road run on 13 November 1896. This car was the first ever to use an electric self-starter for the engine; up until then the engine had to be cranked into life by hand. The starter was not a standard fitting, though it was added by the car's first owner H. J. Dowsing who took out the first self-starter patent in 1896. In fact it was not until 1902 that the first production car with an electrical self-starter as standard equipment was put on sale. This was the Belgian Dechamps car of 1902.

Some inventors wanted more than just an electrical starter, they aimed for a vehicle that was *all-electric* and in 1891 the first electrically driven car was produced. Called the *Electobat* it was made by Morris and Salom of Philadelphia, USA. But the short running times offered by storage batteries meant that the electric car was simply not able to compete with its internal combustion engined rivals. The first of these in the States took to the road in 1893. It was the handiwork of Charles E. Duryea, but although a worthy piece of construction it lacked an efficient speed control system and was swiftly replaced by a more friendly model. This new friendly Duryea went on to win many races, completely under control. But it never won a great market: that prize was reserved for the marque that first ran in 1896, in Detroit. It was a crude home-made contraption, first driven by its maker, whose name was Henry Ford.

For some years the lure of the motor car was felt strongest in Europe. In 1895 there were enough car devotees to stage the world's first long-distance motor race. Organized by *Le Petit Journal* it involved fifteen petrol-engined cars, six steam cars and one electric car. The route was set at Paris to Bordeaux with return to the capital, a distance of some 732 miles. The winner was technically Emile Levassor in his Panhard-Levassor, who made it in forty-eight hours, forty-two minutes. But an objection was raised. His car did not meet the specification laid down by the organizers, it only had two seats!

The car of this period still used many of the features found in the horse-drawn carriage. Wheels had wooden spokes and wooden rims. Large leaf-springs absorbed the shock of bumpy roads. The bodies were coach-built of wood. And passengers and drivers sat out in the open. Some coach builders did offer to make shielding canopies for the travellers: but the first car with a totally enclosed body was not put on sale until 1898. This was a French Renault; and its enclosed pattern was meant to tempt doctors to buy and use the car on their house-calls. It was on the small side but quite nippy at 25mph. Saloon cars took a little longer to appear. The first Saloon model, a Duryea machine, was put on show at the Stanley Motor Show of 16 January 1903. Others were quick to see the appeal of this style and before long the motor car began to loose its horse-and-buggy aspects. The wheelbases became longer, the centre of gravity was lowered and the wheels shrunk in size. Rubber tyres, solid or pneumatic, helped smooth the ride and grip the road. Many swore by the solid rubber tyre, it seemed to offer dependability: no blow-outs; no grimy wheel-changing or puncture-repairing. But as the pneumatics became more dependable, so they came to oust the solids. Yet it was no overnight victory.

The pneumatic tyre went back some years, it was first patented on 10 December 1845 (No. 10,990); its patentee was Robert Thompson, who had the tyres fitted to his own brougham. These first pneumatics were made by Whitehurst & Co. under licence. A second licencee, May and Jacobs, supplied the first tyres to the nobility and for a while the Duke of Northumberland rode on air. Yet despite such aristocratic approval the venture failed to pay and Thompson reverted to solid rubber tyres. It was a failure once more. Only the hand barrows (used by railway porters) had them fitted for a very short period. The problem lay in the lack of a reliable method of toughening the rubber.

Vulcanization of rubber had first been attempted by Jan Van Geuns, in Holland, in 1836 and by Nathaniel Hayward (USA) in 1838, but it was Charles Goodyear who developed the first practical method in 1839 (No. 3,633, 1844). For years, though, the process was still highly experimental and all the rubber used by Thompson stayed far too soft and in hot weather became quite tacky.

At such a stage it is obvious why the more tricky pneumatics would be overlooked in favour of the more stable solids. With tackiness overcome, hard rubber tyres were gradually introduced for hansom cabs, and they proved a winner with the bump-conscious passengers. Pneumatics, by contrast, were relegated to the more lowly status of bicycle and tricycle tyres when they were reinvented by an Irish vet, John B. Dunlop. Dunlop's first tyres were meant only for his son's comfort, but after fitting up Junior's tricycle he saw the future unfold. Those first tyres were stretched into place on 28 February 1888. In June 1888 he fitted improved tyres to a full-size bike for the first time.

This Eldin machine rode so well that Eldin contracted with Dunlop, and had their new racing safeties fitted with his tyres. They went on sale on 19 December 1888. This first commercial machine created such a sensation that Dunlop opened up his first factory on 18 November 1889. Other manufacturers looked on with envy but Dunlop held the unchallenged manufacturing rights, secured by his first patent of 31 October 1888 (No. 10,607). For two years Dunlop was secure, then someone discovered the earlier patent granted to Thompson. From then on, other makers could cater for the growing market for pneumatics. Their first customers though, were carriage makers. In England, Kingston Welch pneumatics were so used in 1892. In France, Parisian fiacres were fitted with Michelin pneumatics in 1894. A year on, and on 11 June 1895 the first pneumatic-tyred motor car took to the French roads. It was Edouard Michelin's Peugeot, as entered in the Paris–Bordeaux race. It only came ninth, since twenty-two inner tube changes were needed, but the ride was smooth.

Car exteriors were naturally the most obvious evidence of new trends; not so obvious, but much more important, were the changes in fuel delivery and mixing; in ignition methods; in transmission systems and in braking. The first disk brake was patented by Frederick Lanchester in 1902 (No. 26,407), though most cars stayed with brake shoes, based on those used on carts. Lanchester also made the first attempt to improve carburation by including a dirt-excluding wick in the carburettor chamber. Until then the carburettor was little more than a warm plate which received drops of fuel and vaporized them. But it was left to Crossley to develop the first effective carburettor (1904) where the mixture was regulated by the suction of the engine and by suction alone. This is the principle used in all modern engines. Employing this principle gave the petrol engine the flexibility and ease of response that made it the certain winner against steam; and electric prime-movers.

At this period the motor car still carried many features taken from coaching days; but gradually new fittings and shapes began to give a distinctive personality. Bumpers came in during the summer of 1905. The first were a pneumatic design manufactured by the Simms Manufacturing Company of Kilburn, London. Electric lamps were commercially introduced in 1908 by the Pockley Automobile Electric Lighting Syndicate in Birmingham. The first Pockley set included headlights, side-lights and a tail lamp. These were meant to be powered by an eight-volt accu-

Frederick Lanchester's highly original car of 1895

mulator. And at this time large steering wheels ousted the antique tiller-steering that had served well for so long. While increased speeds required lengthened wheelbases; wheels became smaller and the centre of gravity lower – all in the interest of comfort and safety. But comfort ended at waist level for most drivers. Motoring goggles, motoring coats, scarves and some type of headgear were essential equipment: since most cars were defiantly open to the wind, dust and rain. The fully enclosed body, though first used by Renault in 1898, took decades before it became the most popular form. Even at 100mph open cockpits were the rule. The first man to reach this speed ended his feat with a face raw from the sand-blast effect of grit and dust. This was the Frenchman Louis Rigolly who drove his 100hp Gobron-Brillie over a timed kilometre in July 1904. His actual speed was 103.56mph.

All of these vehicles were expertly hand-crafted. The very idea of mass-production was anathema to the industry for many years. And yet the production of interchangeable parts had been established as a valuable technique at the end of the eighteenth century. The first man to show its commercial value was Eli Whitney, the inventor of the first cotton gin (1793). But others before Whitney had argued the case for interchangeability; in France artillery General Gribeauval had suggested standardizing cannon carriages in 1756, and in 1785 the French gunsmith Le Blanc made standardized parts for guns, yet these early endeavours had little impact. But when the American Minister to France, Thomas Jefferson, visited Le Blanc's workshop he was so impressed that he sent a full report to his government. He wrote that he had been handed a box holding the parts of fifty musket locks: 'I put several of them together myself taking pieces at hazard, as they came to hand, and they fitted in the most perfect fashion.' These words helped create a favourable attitude towards Whitney when he reached the stage of contracting with the US Government for an order for muskets.

At that time the United States suffered from a shortage of skilled machinists and Whitney saw a way to circumvent this lack. He invented a manufacturing system that would allow an unskilled man to turn out a product equal to that of a skilled craftsman. All the skilled craftsmanship would be confined to the initial stage of making the master-patterns and machines needed for production. Thus the Whitney musket was so designed that each part was embodied in a master template, the very principle used in dress-patterns. The unskilled machinist would then take his allotted template to a work table, clamp it on to an unshaped metal plate and use a cutting tool to follow the template outlines. Special rotary cutters would nibble away at the unwanted metal; little skill was needed to keep the cuts accurate.

To control the cutting, Whitney devised the first milling machine worthy of the name. It used a multiple-edged cutting tool carried on a movable work bed driven by a worm gear and screw thread. Such machines proved indispensable for all future mass-production methods. He added other semi-automatic machines: one to shape and bore the barrels; another to shape the stocks; then began the job of supplying 10,000 near-identical muskets to meet the Government contracts.

From a slow start he increased the pace of production to the point where a grateful government gave him a vote of confidence and a further order for 15,000 guns. It is fair to say that Whitney was not a lone genius, as often depicted, but the first man to gather together all the many ideas around at that time and bring them to a resting place in his manufactory.

Whitney's example influenced other mass-production devices. One of great importance was the copying lathe made by Thomas Blanchard in 1818. This was the first lathe able to turn out *irregular* shapes and Blanchard designed it with wooden gunstocks in mind. He took a pattern of the shape to be turned and arranged for every part in turn to be brought in contact with a small friction wheel. This wheel controlled the position of a cutting tool set to work on a rotating, rough wooden blank. At each rotation the tool pared away all the unwanted wood and duplicated the shape traversed by the wheel. After sandpapering, an exact copy of the original pattern was ready for fitting and varnishing.

Whitney's enterprise soon influenced other industries. First to adapt his ideas to a very different style of work was Eli Terry a clockmaker, settled at Waterbury, Connecticut. Terry installed machinery for clock making in an old mill. His machines turned out interchangeable parts for clocks at such a rate that a production cycle would yield a hundred clocks in the time it formerly took to fashion a score or less. He sold these first mass-produced clocks himself; ranging around the countryside on horseback to tout for customers. In 1807 he undertook to turn out 4,000 clock movements and completed the contract in three years. And in 1814 he invented the first 30-day shelf clock, which revolutionized the clock trade. All these clocks were made from wooden parts.

In 1812 one of Terry's ex-workmen Chauncey Jerome, became the first to break with the tradition of wooden clock making when he switched over to a brass one-day timepiece. The only wood used by him was for a carrying case to contain the clock. His methods became so precise and organized that he reached the point where three of his men could make and cut all the wheels for 500 movements in one day. If we concede that the *value* of mass-production had been well demonstrated long before the motor car arrived; we also have to recognize that it was only applied to small products. Nothing as large as an automobile engine or an automobile itself had ever been attempted. The conveyor-belt system had been foreshadowed, though, by the Chicago meat-packing factories of the 1860s. There, pig carcasses were moved from one end of the factory to the other, on an overhead conveyor system, allowing workers along the route to make their own specialized cuts into flesh and bone.

The first moves towards the mass-production of cars were made by the French, like Renault and Panhard. But mass-production proper was first organized in the USA by the Detroit company Ransome Olds in April 1901. Olds adopted a simple style that lent itself to production at a rate of ten cars a week at the start. This figure jumped, and jumped to the stage where Olds were able to turn out over 5,500 cars in 1904. They were cheap and good hill-climbers, but grossly underpowered. Their engines were made by the Dodge machine shops; their transmissions were made by Lelands; other smaller shops delivered their own bits and pieces, and from all this came the 'Merry Oldsmobile'.

Lelands were responsible for the smooth performance given by the Oldsmobile, since they were master machinists working to such high standards that it was said of them that precision was their religion. In that case they created their own church by breaking with Olds and setting up Cadillac. The supreme article of faith now became total interchangeability of parts throughout.

Cadillac was the first firm to achieve this before it sold out to General Motors in 1909. And it was the first firm to establish mechanical standards for the whole industry. Even so it was not until 1913 that the first modern assembly-line plant took shape. This was Henry Ford's first attempt at the system and in 1914 he became the first manufacturer to adopt the system in all his plants. From such plants poured out the celebrated Model T Ford.

Though first made on 1 October 1908, it was an ideal car for the new assembly lines. It was made as simple as possible, even though it embodied advanced ideas. It used a flywheel magneto to give it starting power independent of a battery. Its engine was a four-cylinder side valve model, made as a single block with a detachable cylinder head. A two-speed epicyclic gear box and a vanadium steel crankshaft took care of the power from the engine. Everything was aimed at mass appeal and it became the first car to sell over a million. Then, in 1923 it became the first car to sell over a million and a half in one year (1.8 million in fact) and by 1927 some 15 million Model Ts had been made, all of them amazingly similar in most essentials.

What made the modern motor car possible, were the many important technological advances all coming together at the same time. The turret lathe, first built by Stephen Fitch of Middlesex, Connecticut, in 1845, gave the lathe operator the choice of eight spindle-mounted tools, which could be brought in use by a capstan advance and engaged feed. In that way eight successive operations could be performed rapidly without having to stop the lathe to change tools by hand. This had been hailed as the most significant development to the lathe since Maudslay made his first classic screw-cutting lathe in 1798. It has even been said of Fitch that 'One only does justice to [his] memory when one recognizes in him the creator of one of the most time-saving machines ever invented.'

Another great time-saver and aid to continuous production was the collet chuck patented by Edward G. Parkhurst (USA) in August 1871. This enabled bar stock to be fed through the hollow spindle of the lathe until the needed length was reached. Then the collet chuck gripped tight, and machining began. At its finish the completed part was turned off, the chuck relaxed and a fresh section of rod was fed through to the right length; the cycle then recommenced. At no time was it necessary to stop the machine.

By combining the collet chuck with the turret head, Christopher M. Spencer was able to create the first automatic turret lathe for turning metal screws. He controlled all the actions of this lathe by means of what he termed a *Brain Wheel*. This was a large drum with steel cams bolted around its periphery. When this drum revolved each cam activated a predetermined part of the turning sequence. When Spencer's first US patent was taken out in 1873 it was seen that he had improved the original by using two *Brain Wheels*, to control production; one for each end of the lathe.

Another great step was taken when Joseph Brown turned his attention to milling machines. He had exacting standards in mind, for his training had been that of a precision instrument-maker. His eagle eye and patience led him to make the first universal milling machine in 1861. It was an inspired design, but at first sight deceptive in its clean, simple lines. When it was delivered to its first owners, the Providence Tool Company (14 March 1862), it perfectly and swiftly solved a major problem that had been troubling the company. It was now possible to turn out the first precision-twist drills and combined with great accuracy was a fast rate of manufacture. Prior to this, all the flutes in the twist drill (first used 1860) had to be made by hand filing, an expensive and slow business.

Then we had the introduction of the first Universal Grinding Machine in 1876. This too was made by Joseph Brown's company, Brown and Sharpe; and it represented the high point in the development of grinding devices, a development that traced itself back to the first early machines of the 1830s. Those machines used natural sandstones or emery granules and soft wheels to surface finish metal work, but only a high finish could be reached. There was no question of using these machines to turn out work to exact dimensions; for one thing the performances of the abrasives used could not be predicted, so precision was not even aimed for. The first attempts to produce predictability involved bonding crystals of emery or corundum (a form of aluminium oxide) with adhesives to form solid grinding wheels. The first bonded wheels appeared in 1840. Then the first attempt to produce a vitrified wheel was made in 1842 by Henry Barclay. His baked clay and emery wheels were on the right track but lost strength and shape when fired. It was not until 1873 that Sven Pulson, a potter at Franklin Norton's workshop in Worcester, Massachusetts, made the first successful artificial grinding wheel. He too used clay, like Barclay; but by adding *clay slip* (a glazing mix) he obtained stability. Pulson's method, first patented by Norton in 1877, made it possible to begin precision grinding. It also led to the first firms devoted to making grinding wheels, and to J. M. Poole's heavy roll grinding machine of 1870. This worked on principles never used before in a machine tool (the principles are still in use today) and for the first time ever allowed a degree of precision in the finished work never achieved before, except on small pieces of work. It was accurate to within 0.000025 of an inch!

In 1891 came a great discovery by Edward G. Acheson (USA). He fused a mixture of carbon and clay in an electric furnace and created crystals second only to diamonds in hardness. He had discovered silicon carbide, which he promptly dubbed 'carborundum'. Now, though he did not know this, early that century someone else had hit upon the same substance, but just regarded it as a curiosity, and with good reason. There was then no way of exploiting the find. Acheson's re-discovery as we could term it came at precisely the optimum moment. To make the stuff on a commercial scale demanded an electric furnace, and there, just about to open up at the Niagara Falls, was a huge hydro-electric generating plant. So the logical place for the world's first carborundum factory was obviously at Niagara Falls.

In 1901 the Norton Emery Wheel Company joined the Acheson Carborundum Company at Niagara Falls. Norton, though, was there to manufacture a rival grinding wheel based on the first process able to make an artificial corundum. This was Charles Jacob's process, first developed in 1897, in which aluminium oxide was fused with coke and iron borings. Another Norton, unconnected with the Emery Company, experimented along unorthodox lines and concluded that *wide* grinding wheels, perhaps up to a foot or more wide, could turn out precision work and even handle rough work.

At that time only near-finished work was ever sent to the grinder, so Charles Norton's views were workshop sacrilege. As well as that the grinding wheels were always trued to a narrow cutting edge, so the idea of a broad wheel-edge was just the raving of an idiot. This reactionary attitude towards his views led Norton to leave Brown and Sharpe in 1899 and set up on his own. An old friend, Charles Allen, helped the new firm to get on its feet, and by March 1900, Charles Norton had designed his first heavy production grinding machine. By November he had produced the first two machines and the first lucky company to buy one found themselves owning a machine that lasted them for thirty years.

Norton's design established the principle that the key to precision is *rigidity*. He also first proved that it was possible to contour the edges of his broad wheels so that they ground the required shape into the work in front of them. In 1903 he used this technique to make the first special crankshaft-grinding machine. One of his contoured, wide grinding wheels, could produce a finished crankshaft in fifteen minutes. The alternative method of turning, filing and polishing took at least five hours. The first company to buy a Norton crank-grinder was Locomobile US. Of his work it has been written, 'His machine represents an historic milestone of the greatest significance for from it there quickly stemmed all the special-purpose production grinding machines which alone made "motoring for the million" possible.'

Other, less obvious developments came to make motoring popular. Ball-bearings were perfected when Hiram Baker and Francis Holt of Manchester created the first ball-grinding machine in 1853. In the 1880s the next step was taken when the first machine for grinding *hardened* metal balls was made in Coventry. At first only the smaller balls meant for cycle bearings were ground; but the technique had been established and was there for expansion. It was a technique easily adapted for roller bearings and at the turn of the century Frederick Lanchester became the first motor manufacturer to pioneer the use of roller bearings in the industry.

By 1930 every manufacturer regarded ball and roller bearings as essential, but when Lanchester needed roller bearings for his first epicycle gear-box and rear axle, no manufacturer of ball-bearings would take his order. He had to make his first set of roller bearings with his own hands, grinding the steel rollers six at a time, to a tolerance of 0.0002 an inch. His grit and enterprise made Lanchester the first man to make a truly British car, and the first in England to adopt the principle of complete interchangeability of parts in his 10hp car of 1903. This used over a thousand parts, many of them made to the unilateral system first devised by him in 1895. Unilateralism rejected the American bilateral approach, which gave every precision-made part a plus and minus toleration for each dimension. Lanchester's system was simpler: his dimensions were given as the ideal fit with minimum clearance; a single tolerance figure indicated the maximum permissible machining error. This system eliminated interference fits in his machine shops, to the benefit of both maker and customer.

Another behind-the-scenes factor lay in the introduction of high-speed steels for cutting tools and the improvement of internal grinding methods. In large-size engines, like those for steam, the grinding of cylinders presented little problems. When dealing with a small petrol engine, with its multi-cylindered block and its narrow thin-walled bores, there arose problems of deflection of reamers by the inevitable soft and hard spots in the casting.

In 1905 James Heald (USA) overcame this difficulty with the first planetary grinding machine. He made his grinding wheel revolve at a high speed and made the wheel spindle revolve as well, but at a slower speed. This involved using eccentric bushing and eccentric spindle mounting; all subject to fine adjustment. With this arrangement the grinding wheel only exerted a very light pressure and was not deflected by hard or soft spots. His first machine of 1905 took rough bored cylinders and finished them perfectly straight and parallel to within 0.00025 of an inch. The machines used today all work on these same vital principles, but Heald's name is scarcely remembered.

In fact few of the pioneering names are remembered; we tend to recall Ford and Morris, Rolls and Royce and hardly anybody else. The marque names, though, are still chewed over wherever car fanciers

The first Rolls-Royce

gather. In all over 180 marques have existed since the first car took to the road.

However you will not find the name 'Selden' in that list and yet it is to George B. Selden (USA) that the honour falls of being the first to apply for a patent on a 'horseless carriage' driven by an internal combustion engine, fuelled by a liquid hydrocarbon, which as a matter of course, covers both kerosene and gasoline. His patent was first applied for on 8 May 1879, and at that date was a highly original and ingenious concept. But Selden lacked the skill to make the plans come to life; so he stalled for years holding back the granting of the patent by filing amendments. This was quite legal under US patent laws and, as a result, this first patent was not granted until November 1895. After that he sold out his rights to a syndicate which became the Association of Licensed Automobile Manufacturers. This body then used the patent to extract revenue from all manufacturers of motor cars in the States. Ford and a few others refused to pay up and the patent became the pivot of see-saw battles in the courts. This was the first legal battle occasioned by the automobile. For the record, it should be noted that in 1905–6 the US courts demanded an example of Selden's car, as laid down in his specifications. His son reluctantly set to work and after much effort produced a vehicle. He even painted the date '1877' on the body to make it look more impressive. When it was unloaded for its witnessed trial it looked good, even if it had the air of an antique. But when the first Selden car based on those first patents made its first run it covered a mere 1,500 feet before it conked out. The court was no longer impressed. That his side won the case was due to a technical knock-out, more than the triumph of right. And Selden can be rightly summed up as an idle dreamer intoxicated by the fumes of the first phantom car.

Phantom car fumes are indeed the stuff of dreams, but even while the Selden battle was raging there were practical idealists who were trying to deal with the worst aspects of exhaust fumes. On 17 April 1909 Michel Frenkel of France patented the first method of scrubbing and deodorizing exhaust fumes. That it had little effect on manufacturers shows that the problem had not yet become great enough for popular discontent and pressure to be noticeable. In fact it took a further sixty-three years before the first successful exhaust fume cleaner was developed. In 1974 General Motors (USA) introduced their catalytic converter, an excellent solution to the problem. This converter was designed as part of the exhaust system, not a bolt-on accessory. It worked by passing the fumes through a chamber packed with small ceramic pellets each one of which was coated with a precious metal. Platinum or palladium or rhodium could be used. The metal layer acted as a catalyst; that is, it speeded up the action without itself changing; and the toxic substances were then converted into less harmful substances. These by-products were speedily absorbed by the ceramic cores which were made like miniature honeycombs. And a 'clean' gas was then ejected in the normal way. This cleaner system became compulsory first in the USA and then in Switzerland in October 1987.

Many other countries are moving in the same direction, and in 1988 Ford made its own contribution to this healthy trend by producing the first platinum-free catalytic exhaust. By dispensing with the precious metal

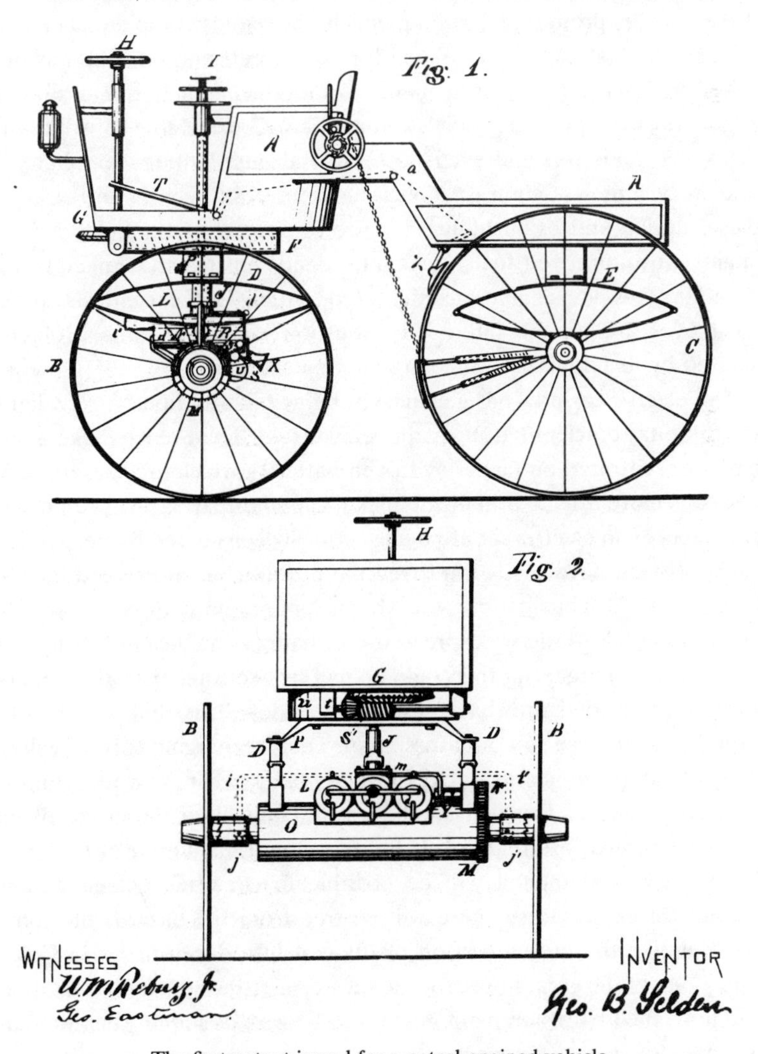

The first patent issued for a petrol-engined vehicle

they have slashed the price of the system and made it more attractive. A different solution was offered by Toyota of Japan, also in 1988. At the Geneva Motor Show in March they unveiled their first 'low-pollution engine'. It met the prevailing anti-pollution standards without recourse to a catalytic exhaust. Their example was followed by DAF of Holland who produced their first 'clean engine' in 1990. Their approach differs from that of Toyota and DAF even claims that no other system will rival theirs for cleanliness.

There still remains the problem of fuels that are inherently dirty. The first clean-up move was made in 1975 when unleaded petrol was first put on sale in the USA. But the new fuel was slow in being accepted, indeed the first unleaded petrol was not sold in Britain until January 1986. Yet there are very promising developments in the realms of the diesel engine. In 1980 the first patent was issued for the Elsbett engine. This was first tested in 1984 and ran on a modified rape-seed oil. It is a multi-fuel engine operating on direct injection and is said to be able to work with many vegetable oils and even waste animal fats. Its pollution factor is said to be tiny. By contrast Volkswagen have stuck with an orthodox diesel design and by including a turbo-compressor have made a 60hp engine with improved fuel consumption and thus cut the emission-rate per mile. This new engine was first fitted to their Golf car in 1989. While most eyes are on the internal combustion engine and the problems created by its popularity, some are looking elsewhere.

The electric car now has a chance of being taken seriously. New lightweight, ultra-efficient batteries are being tested. The French makers Saft have concentrated on nickel-hydrogen batteries which are said to be 50 per cent more effective than the nickel-cadmium rivals. At present the prospects of an electric car are being seriously considered by Peugeot. In Japan, Nissan have vastly improved the nickel-iron battery and in 1983 they produced their first car, the Micra, running on battery power. The main drawback to the widespread use of battery cars lies in the need to constantly recharge. But this could be met by exchange batteries held by garages. A second drawback, usually overlooked, is that battery cars would not conserve stocks of fossil fuel. Their recharging current will be drawn from power stations that for the greater part now work by burning petroleum oil. This will all change dramatically if the hopes of the nuclear scientists are realized. If the thermonuclear *fusion* of hydrogen can be achieved then it will be possible to construct power stations around fusion reactors. These will be free from the hazards of atomic waste unlike the nuclear fission plants first introduced in 1951. This is surely one of the great hopes for the future, and since the hunt is up, here and now, the first fusion plant might well be with us sooner than we dare imagine. So, watch this space . . .

• AVIATION •

LIGHTER THAN AIR

To fly like a bird was one of the great dreams of early Man. It inspired many a dreamer to strap on grotesque wings of fabric and wood and dive from towers or cliff tops. The first of these recorded jumps was made by King Bladud, the ninth British monarch, in 843 BC. Neither his rank, nor his superior wings of feathers, were sufficient to save him from a messy death.

When Man first rose into the air it was as a passenger on giant kites. Marco Polo gives us the first account of such a flight in fourteenth-century China. From then, until the age of the balloon, only model aircraft provide any sort of aerial variety.

The idea of using heated air as a lifting agent led to the first hot-air balloons in the eighteenth century. On 8 August 1709 Father de Gusmão demonstrated his model balloon in Lisbon indoors in the presence of King John V of Portugal. It rose a full 12 feet, endangered the expensive curtains with its fire-box, and was promptly shot down by grooms.

After that, most demonstrations took place outdoors and on 25 April 1783, the Montgolfier brothers (Joseph and Etienne) released the first hot-air balloon capable of carrying a man. It rose to some 1,000 feet (305m) and travelled some 3,000 feet (915m) away from its lift-off site. Its 39 foot-diameter envelope was filled from hot air rising from the wood-shavings and straw that burned beneath the balloon's neck.

Father Francesco Lana's 'flying boat' of 1670. Filled with gas, his globes could have provided lift, but he wanted them pumped free of air; that meant total collapse

The Montgolfiers' rival, Jacques Alexandre César Charles, scorned the use of hot air, and on 27 August 1783 he launched the first balloon filled with hydrogen. It drifted 15 miles from Paris to Gonesse, where it was hacked and slashed to pieces by frightened, superstitious peasants. It expired with much wind and hiss.

The Montgolfiers responded by sending up the first living creatures on 19 September 1783. Their balloon carrying a duck, a cock and a sheep, descended after a two-mile flight, with its three passengers intact and unruffled. They quickly followed this with the launch of a balloon carrying the first man aloft. The world's first aeronaut was François Pilâtre de Rozier who ascended and descended on 15 October 1783. It was a tethered flight only, but it led to the first free flight on 21 November 1783. The passengers were de Rozier, once more, and the Marquis d'Arlandes. Their flight carried them some 5.5 miles, over and beyond Paris, in 25 minutes. Even so, hot air, as a lifting source, was grossly inferior to Hydrogen. On 1 December 1783 Jacques Charles proved this when he took aloft one of the Robert brothers (makers of rubber-coated silk). Charles' treated-silk balloon of 28-foot diameter, landed 27 miles away from its start in Paris.

After that, few doubted that the gas-filled envelope was the winner and on 15 September 1784 an Italian brought the French invention to London and made the first British aerial voyage in a hydrogen balloon. Vincenzo Lunardi's voyage took him from the centre of London to Ware, in Hertfordshire. A stone monument still marks the landing point and records 'That wonderous enterprize, successfully acheived By the powers of Chymistry And the fortitude of man . . .' Then, within months, came the first aerial crossing of the formidable English Channel. On 7 January Frenchman Jean-Pierre Blanchard and his American friend Doctor John Jeffries took off from Dover and landed in the Forêt de Felmores, in France, just two-and-a-half hours later.

Sadly our first ballooning aeronaut, de Rozier, met his end in the first fatal accident involving balloon flight. Along with Jules Romain he died in the first attempt to fly across the English Channel from the French side, on 16 June 1785. An act of sheer folly had led him to add a Montgolfier hot-air unit to his hydrogen balloon. When the hydrogen vented, his craft exploded. He never even made it as far as the water's edge.

Ballooning reached the New World on 9 January 1793, when the French pioneer Blanchard made the first free flight in America. He took off from the old Walnut Street Prison yard in Philadelphia and flew for 46 minutes. It was the first flight to be witnessed by the first President of the USA, George Washington. And records show that John Adams, Thomas Jefferson, James Madison and James Monroe, all future presidents, were there as well.

Blanchard made thousands balloon-conscious, until a heart attack while aloft ended his career. His widow took over his career and became the first woman aeronaut to die in a flying accident. Her hydrogen balloon was ignited by a stray firework at a display in the Tivoli Gardens in Paris. The date was 7 July 1819.

Military use of the Balloon began as early as 26 June 1794, when the French Republican Army sent up their first reconnaissance balloon *L'Entreprenant*, to observe the enemy at the battle of Fleurus. And it was the French who created the world's first 'air force' when they formed the Aérostatic Corps of the Artillery Service on 29 March 1794. The novelty of the Aérostiers led to much excitement among the fair sex. Something akin to modern day pop-groupie cults grew around them and there were reports of assignations in the air and bouncing balloon baskets. Wilfrid de Fonvielle reported that 'The favour of the ladies followed the balloonists wherever they went, which was not an unmixed blessing, and seems in the end to have contributed to the suppression of the corps.' So this must be the first time that an air force was defeated by feminine wiles. It was a short life and a frilly one. And while it lasted, all its duties were genteel. Aérostiers simply watched and reported, bombing was out of the question.

But inevitably bombs were called for and the first balloon-borne bombs were despatched by the Austrians against the defenders of Venice on 22 August 1849. They were unmanned, hot-air devices carrying 30lb bombs. They caused little damage and much amusement. The first regular Balloon Corps did not come into being until 1 October 1861. This American Army Balloon Corps was furnished with fifty men, seven balloons and commanded by Chief Aeronaut Thaddeus Lowe. Before taking up that post Lowe had installed the first telegraph apparatus in the basket of a balloon and with it he sent the first telegram from the air. It was directed to President Lincoln on 18 June 1861. It read, in part, '... I have pleasure in sending you this first despatch ever telegraphed from an aerial station, and in acknowledging indebtedness for your encouragement for the opportunity of demonstrating the availability of the science of aeronautics in the military service of the country.'

The American Civil War also saw the construction of the first aircraft carrier, when the barge *G. W. Parke Custis* was rebuilt as a mobile balloon-launching and towing vessel. It entered service in November 1861.

Nine years on, during the Siege of Paris, came the first airmail service. The Prussian troops had the French capital surrounded, with all its land communication lines severed. Then, on 23 September 1870 the French broke the communications blockade by sending aloft a balloon carrying 227lbs of mail. It shot over and beyond the Prussian lines and brought news to the outside world and the provisional government at Tours. The balloon was in the capable hands of Jules Durouf, the world's first aerial postmaster. This triumph led the postal department to set up the first factories for the fast and large-scale construction of postal balloons.

All these balloon trips were one-way only, so to cater for return mail the post office set up the first official pigeon-post and the first microfilm unit. The photographer Dagron perfected a micro-photograph technique which could cram almost a thousand pages of letters on to an ultra-thin strip of collodion. Six of these strips at a time could be sent back into Paris by pigeon, where they were then projected by magic lanterns, copied down by clerks and sent through the regular city post. To combat this audacious balloon-post Krupps introduced the first anti-aircraft gun, so the French switched to night flights.

Before the Franco-Prussian War ended the French had dropped the first aerial propaganda leaflets and issued the first aerial newspaper, the lightweight *Balloon Post.* In all we should allot them the extra credit of organizing the world's first air-lift. In 1883 Britain advanced ballooning techniques with the introduction of Templer's lightweight envelopes made from goldbeater's skin. A year later Britain produced the first gas-tight storage cylinders for the essential hydrogen. Yet despite all advances the great drawback of the balloon lay in its dependence on air currents for travel. It could not be steered or driven to exact destinations. These limitations led to the search for the dirigible, or steerable, airship.

Early thoughts on dirigibles led to schemes for sails or wagging wings. But the first published design of note dates back to 1784, when an officer of the French Corps of Engineers submitted a plan to the Académie des Sciences. Lieutenant Meusenier's plan showed three large two-bladed propellers as the driving force. Eighty crew members would be needed to turn the shafts linked to the props. But such a battalion of toilers was impracticable and the first manned dirigible flight was made by one man controlling a steam-engine. That man was Frenchman Henri Giffard, who took to the air on 24 September 1852 and with his 3hp engine driving a single propeller, steered his airship from Paris to Trappes, a distance of 17 miles taken at a speed of 5mph. But it was an under-powered craft only able to operate in extremely calm weather. Giffard realized this, knew his grander designs would never be financed, and dropped out of the reckoning.

It took another twenty years before an internal combustion engine was first fitted to an airship. On 13 December 1872 Haenlein put his airship on show. It was based on his patent first applied for in April 1865, but

The war balloons of 1870 produced frustration

although it had years of thought behind it, the showing never went beyond a tethered display. All later developments were shelved by the lack of funds.

Others turned to electrical power-plants for propulsion and two brothers, Gaston and Albert Tissandier developed a dirigible using a battery-driven Siemens electric motor. When tried in 1883 it proved to be no better than Giffard's machine. It lacked power.

The first electric dirigible to prove its worth flew in 1884. It was the joint creation of two French army engineers: Charles Renard and Arthur Krebs. The two had developed lightweight storage batteries and this improved the power-to-weight ratio so much that they were able to make a confident, fully controlled return flight on 9 August. They demonstrated conclusively and for the first time, that given a light enough and powerful enough motor, then controlled flights were possible even in high winds. Yet electrical motors were not to provide the solution. On 12 August 1888 an airship designed by the German doctor Karl Woelfert, was tested with power provided by a Daimler petrol-engine. It was the first use of the vital petrol-engine in aviation, and a revolutionary step forward. No one could have sworn to it that day, but as it turned out, only that type of engine was destined to make all forms of powered aviation advance and triumph.

Parallel with the use of the petrol-engine came important developments in metallurgy. In 1886 P. Héroult in France and C.M. Hall in America both invented the electrolytic process for the production of aluminium. For the first time this lightweight and strong metal could be produced on a large scale at an affordable cost. It ceased to be a precious metal.

Aluminium girders made it possible to construct airships with rigid bodies so that in flight the craft never altered shape. It was this development that led to the first of the world-renowned Zeppelins. Count Ferdinand von Zeppelin had first designed his rigid airship in 1894. The German Government rejected it. He overcame his disappointment and obtained the first patent for his brainchild in 1895. He founded his first company for the Promotion of Airship Flight in 1896. By 1899 he had constructed the first floating dirigible hanger, basing it at Lake Constance. From this hanger, on 2 July 1900, the Count flew in his first Zeppelin LZ1. The airship was powered by two Daimler 16hp engines, so there was no shortage of thrust, but there were problems in controlling craft.

While Zeppelin worked on his problems, fully controlled dirigible flight was first demonstrated by Alberto Santos-Dumont on 19 October 1901, when he flew his airship round the Eiffel Tower and completed all the other stages set out by the terms of the Deutsch Prize for aerial navigation. He won the 100,000 francs and the government of his own country, Brazil, added a further 125,000 francs and a large gold medal. Even so his feat was a touch-and-go affair and his craft was not suitable for heavy weather or long flights. The first fully controlled air journey was made by his rivals, the Lebaudy brothers, on 12 November 1903. Their 37-mile flight from Moisson to Paris showed that they had designed the first practical dirigible of all time. Their designs were the first adopted by the French Army.

The German Army was equally air-conscious and chose the Zeppelin LZ3 (first flown on 9 October 1906) as its first air-arm, on 20 June 1909, restyling it Z-1.

The British Army had developed and flown its first dirigible somewhat earlier, on 10 September 1907. This ship, No. 1, *Nulli Secundus*, had a three-man crew which included Samuel Cody, soon to become the first man to fly an aeroplane in the British Isles. But it was a civilian dirigible that made the first leaflet raid on London. On 21 October 1908 a hired craft flew over the House of Commons, while a suffragette tossed out leaflets demanding 'Votes For Women!'.

The inevitable first commercial airline came into being on 16 October 1909. This was Count Zeppelin's *Deutsche Luftschiffahrt Aktiengesellschaft.* As *Delag* it flew between the major German cities and on 19 September 1912 it started the first international airship service when it opened its routes from Hamburg to Copenhagen and to Malmö in Sweden.

The First World War ended civilian flights, the Zeppelin now became a relentless war hawk. On 19 January 1915, three German Navy Zeppelins were sent to stage the first aerial bombing raid on Britain. L-6 had to return to base but L-3 bombed the Great Yarmouth area, while L-4 hit seven other areas. Four people died in these raids; they were the first war casualties on British soil. The first air raid on London followed on 31 May 1915 and the raiders escaped intact. The first Zeppelin to be downed was the LZ37, which was intercepted over Ghent on 7 June 1915. This first Zeppelin kill was the work of Sub Lieutenant Warneford of No. 1 Squadron Royal Naval Air Service. He took his plane over and above the airship and unloaded six 20lb bombs. One hit and LZ37 was no more. The action won him a Victoria Cross.

Over British soil the first airship kill took place on 2 September 1916. A Shutte-Lanz airship No. SL X1, was shot down near Cuffley in Hertfordshire. It was the first victim of the newly-invented Pomeroy incendiary bullets used in the guns of Lieutenant W. Leefe Robinson's plane. Another VC, this time for the Royal Flying Corps.

When peace returned the first airship crossing of the Atlantic was made in July 1919. The British airship R-34 left Scotland on 2 July and reached New York on 6 July. It then left New York on 9 July and landed in Pulham, Norfolk on 13 July. Two firsts.

Up until 1921 all dirigibles depended on hydrogen to provide lift. Yet its snags were obvious. It was highly inflammable and an ever present source of danger. On 1 December 1921 the US Navy *Goodyear* airship took off inflated with helium gas. Helium had less lifting power than hydrogen but it could not explode or burn. It was the safety gas aeronauts had dreamed of, but it was then found only in the USA, and for military reasons, its use was denied to all other countries. This denial led to the great airship disaster that hit the *Hindenburg* in May 1937 and killed off the Zeppelins. But before that tragedy, the *Hindenburg*'s sister ship, *Graf Zeppelin* had made the first airship trip around the world. It left New Jersey on 8 August 1929, circled the Earth and returned to its starting point twenty-one days and five hours later. This, the most famous of them all, was scrapped in 1940 and the reign of the giant airships came to an end.

Still the idea of lighter-than-air craft still lives on and the first designs for sky-cranes and flying cargo-warehouses are now nearing their nascent time.

HEAVIER THAN AIR

The father of the aeroplane is now universally acknowledged to be Sir George Cayley (1773–1857) the Yorkshire Baronet, who lived at the family seat of Brompton, near Scarborough. His interest in aviation began after reading accounts of the first toy helicopters made by Launoy and Bienvenu in 1784. These were little more than rotating sticks with vanes at both ends. When spun fast they rose briefly into the air. But this trivial plaything fired his imagination so much that by 1799 he produced the first design for a fixed-wing aeroplane. In 1804 he made his first model plane. It was in glider form, yet it embodied for the first time the essentials of the modern aeroplane: wings, body and tail-unit. Further experiments led to his full-size glider of 1809. When it flew it was able to lift the person launching it from the ground. This was the first flight by a scientifically planned, full-size machine.

Not all Cayley's efforts were devoted to aeronautics, and it was thirty-two years later before he entered on his second fruitful phase with fresh designs, and 1849 before he made his next full-size machine. This was a triplane craft and it was used to take a boy of ten on a short aerial journey. It was the first multi-plane craft to fly, but its design was eclipsed by his next project, a man-carrying glider whose plans were revealed in the *Mechanics' Magazine* on 25 September 1852. Here we have, for the first time, a scheme for a man-carrying craft controlled by its pilot. And within a year, Cayley had actually used his new ideas to build a man-carrying glider which flew across a dale at Brompton, taking a coachman as its trembling and unwilling passenger! This was the first flight of its kind.

Historian Charles Gibbs-Smith has listed twenty-seven firsts for Cayley. Some of these are rather technical, but among them we find the first (1853) use of stretched rubber to power a model aircraft; the first

Sir George Cayley, 'Father of Aviation'

suggestion for using an internal combustion engine for aircraft propulsion and the first design for a tension-spoked wheel. This wheel (of 1808) was intended to be used for aircraft undercarriages and it proves to be the forerunner of the spoked wheels so familiar to us from their use on bicycles.

Strange to relate, Cayley's ideas never made much impact. William Henson's *Aerial Steam Carriage* of 1843 ignored many of Cayley's calculations. When tested, in 1845, it failed to fly, though in theory its 20-foot wingspan was enough to sustain it. Yet it was the first airscrew-driven, fixed-wing design of all time. And it was the first to use twin airscrews; the first to use a wire-braced monoplane construction; the first with a combined elevator and rudder tail-unit; and the first to employ cambered wings built up with shaped ribs. It was let down mainly by its over-weight steam engine. But in those days steam was still the force that first came to the minds of most designers. It was certainly uppermost in the mind of the French experimenter Felix Du Temple. His first model aeroplane took off in 1857 powered by clockwork, but this was soon fitted with steam-propulsion and still proved its worth. It was the first to take off under its own power, fly and then land itself safely. And in fact the first to be tested in French history. This led to his first patent, of 2 May 1857, for a full-size machine, a machine equipped with the first retractable undercarriage. This machine was built, tested, modified and in 1874 became the first powered plane to leave the ground. Yet it only remained airborne for a brief time and even then this feat was assisted by allowing the plane to run down the slope of a take-off ramp. Here we should note that the Russians have always claimed the first powered-flight for themselves. Their claim involves the 65-foot effort made by Alexander Mozhaiski's steam-powered monoplane. Now this monoplane record may be a first for Russia, but it was not made until 1884, ten years *after* the French effort.

Outstanding among the French theorists was Count Ferdinand of d'Esterno. His work on the flight of birds of 1864 inspired others to want to *soar* like birds. And he provided plans for an attractive glider that later strongly influenced the power-plane pioneer Gustave Whitehead. His ideas also influenced the French pioneer Clément Ader. Ader is a figure of much controversy. He developed an excellent light steam-engine delivering between 18 to 20hp. This he saw as the ideal power unit for flight. And his first monoplane the *Éole*, using this engine, certainly took off on 9 October 1890, and covered some 165 feet, with Ader as pilot. But, though a notable first, it was never a sustained or controlled flight. Then, seven years later, Ader made his bid for the great goal of controlled flight, this time with a new plane, his *Avion III*. His attempts spread over three days in October 1897, and on 14 October they ended in failure. Then, astonishingly, in November 1906, Ader claimed that his attempt of 14 October had resulted in an 'uninterrupted flight of 300 metres (985 feet)', and he had witnesses to prove it. It was a claim that led to bitter disputes that dragged on for years, but the last word should go to a fellow-Frenchman Charles Dollfus, Curator of the Paris Musée de l'Air. After studying every scrap of evidence he concluded: 'Ader did not fly for a single instant at Satory during the tests of October 12th and 14th 1897.'

Ader's last work had been commissioned by the French Government, with one eye on future warfare and a similar concern prompted the US Government to take an active interest in aviation. The Spanish-American War of 1898 led to the setting up of a military committee to look into aviation's future role. The man who interested them most was Samuel Pierpoint Langley, mathematician, astronomer and Secretary of the Smithsonian Institution, the USA's great National Museum. Langley had taken up the challenge of flight in 1887. He built a number of steam-powered model aeroplanes and found them all failures. Then he rethought things and rebuilt two of his models. His rebuilt model No.5 brought him success. On 6 May 1896 it flew for 3,300 feet, the first record for the USA. When his No.6 flew for 3,960 feet he grew certain that he was on the right track. All he lacked now was the time and money to build a large man-carrier. A $50,000 subsidy from the War Department give him the impetus to begin work afresh.

He began by commissioning the first US aero-engine from Balzer's of New York. When delivered in 1899 it failed to yield up the 12hp Langley had specified. Charles Manly then redesigned it to meet an even higher specification of 24hp. He exceeded this by making the engine yield up an incredible 52.4hp. This first aero-engine of realistic power was first tested in January 1902.

Before trying for full-scale flight Langley completed a quarter-size model of his project. This first flew in August of 1903. It was a tentative flight, but still the first by a petrol-engined aeroplane. Then his full-size plane was made ready for take-off on 7 October 1903. It was a perilous take-off. Langley had decided to launch the plane from a huge catapult mounted on a houseboat moored on the Potomac river. His pilot was Manly, a brave man without the slightest experience of flying or gliding. Up it went and down it came, flop, into the river. Manly crawled out in one piece, but no one was any the wiser. The very same folly was repeated on 8 December 1903 with the same results. Years later these failures, like those of Ader's, were to lead to bitter battles of words, actions in courts, and unacceptable restrictions on historical research.

While power-flight still remained elusive, gliding flights were providing experience and valuable knowledge of aeronautics. And greatest of all these hardy glider flyers was the German Otto Lilienthal (1849–96). In 1893 he was the first man to take off, fly, control and land using a glider. This first completely reliable glider was based on his own design and built by him. His book *Bird Flight as the Basis of Aviation*, was an invaluable handbook for future aircraft designers, including the Wright brothers and Gustave Whitehead. But he never lived to witness power-flight. He was working on the idea himself when he stalled in his glider, crash-landed, broke his spine and died the next day. His influence has been rightly assessed as 'universal and profound'.

Lilienthal's example and teachings resulted in the first manned and controlled glider flights in Britain. His admirer, Percy Pilcher, made these flights in 1896 in a plane of his own design, which makes it a double-first. What is more, he used a towing technique for take-off for the first time ever. On top of that he patented (No. 9144 of 1896) the world's first practical design for a powered plane. He was still working on his engine to drive such a plane when, like Lilienthal, he crashed while gliding and died of his injuries on 2 October 1899.

In the end neither the French nor the British were able to claim the first sustained and controlled flight of a powered aeroplane. But who really made it first? At first sight it seems that few would challenge the status of the Wright brothers, and yet the Germans now claim the honour for a German mechanic who settled in Bridgeport, Connecticut. This mechanic, Gustave Whitehead (Weisskopf) was certainly devoted to the conquest of the air. He stated that he had been first inspired by watching and working with Lilienthal in Germany. On arriving in America in the 1890s, he began constructing model aircraft and gliders. On his marriage license of 2 November 1897, he even described his occupation as 'Aeronaut', this was at a time when the Wrights were still building bicycles.

Gustave Whitehead. Was he the first?

It is claimed for Whitehead that he first flew a manned aeroplane in April or May 1899. It was steam-powered, but crashed into a building before it was fully tested. No claim has been made, though, to count this as a controlled flight. And Whitehead himself never wrote about it. Whitehead's own claims are much more startling. In 1901 he said that he had flown for half a mile in one of his flying machines, on 14 August. Then, in a letter to the *American Inventor* Vol.IX–No.1, April 1902, he records that on 17 January 1902 his latest aeroplane, No.21, flew nearly 2 miles over Long Island Sound and landed safely in the water. A second flight, made at a height of 200 feet, described a huge circle in the air and covered a round distance of about 7 miles, before setting down on to the water. The plane, it should be noted, was meant to float. Two things should be recognized about this statement: the flight was said to have been *witnessed* and Whitehead did ask the magazine editor to come and photograph his next planned flights. But there is no record of any attempt to interview witnesses, neither did the editor nor any other newspaperman take up the offer to visit and photograph.

Whitehead's record, from then on, is somewhat erratic, but the same can be said for many other inventors, including Cayley. And a growing body of thought in the USA now takes his claims seriously.

By contrast the Wright brothers look very stable, and respectable as one would expect from the sons of a bishop. But Bishop Wright was a man with problems involving bitter fights within his small, rigid evangelical Church of the United Brethren. The brothers grew up believing in the essential depravity of man. Outsiders were folk not to be trusted. This outlook may well have determined their attitudes towards their rivals and to truth itself.

Their claim to the first sustained and controlled powered-flight rests on events that took place on 17 December 1903 at Kitty Hawk, North Carolina. On that day their twin-prop biplane the *Flyer*, had made four ascents. The first was a twelve-second excursion that took the plane a mere 120 feet. The fourth and last trial took the plane 852 feet in fifty-nine seconds. Experts still argue over which flight should be labelled the first. But the Wrights' first flight certainly took place on that day. Yet was it the world's first? Or did shabby, German-tongued Whitehead beat them to it? Sad to say no powerful body in the US ever bothered to check on the Whitehead claims. The Wrights had the monopoly of trust. And the Wrights used their position to virtually outlaw any research or findings that conflicted with their claim to priority. This is best shown by the conditions imposed on the Smithsonian by Orville Wright. That institution had angered Orville when it made extravagant claims for the plane made by its one-time Secretary, Langley. In pique Orville sent the original *Flyer* to the British Science Museum in London. He wrote, '... sending our Kitty Hawk machine to a foreign museum is the only way of correcting the history of the flying machine, which by false and misleading statements has been perverted by the Smithsonian Institution ...' This move hurt badly for the Smithsonian saw the plane as the supreme aviation jewel.

Finally, after twenty years of wrangling the American National Museum gained the *Flyer* on condition that they would never publish or display a statement claiming that any aircraft earlier than the Wright Aeroplane of 1903 was capable of carrying a man under its own power in controlled flight. If that great museum ever acknowledges an earlier claimant, then it automatically loses the Wright *Flyer*. A binding contract dated 23 November 1948 holds this threat over the Smithsonian. And it was implicit for many years before, during the long fight between Orville Wright and the museum authorities. It may have suited both sides, but it acts as an effective gag on truth.

Still, whether Wright or Whitehead, we can say that it was *the Wrights'* efforts that gave the needed boost to all further strides in the air, at that time. Crude gliders, based on the Wright patterns, were first used in France by Captain Ferdinand Ferber, in 1902. This was the first European use also. The first European plans for such gliders were published in an article by Ferber in the February 1903 issue of *L'Aérophile*. When the French first heard of the Wrights' power-flight the air-minded among them echoed these words of Ferber: 'The aeroplane must not be allowed to be perfected in America. There is still time, but let us not lose a minute.' But while they fumed and fiddled away the Wrights perfected the world's first practical powered-plane and gave it its first flight on 23 June 1905. This plane, *Flyer III* was robust, could fly in figures of eight and remain airborne for over thirty minutes. Stubborn to the end, the French refused to believe that the Wrights had triumphed. It was only in January 1906 that it was first conceded that their dreams were over, when pioneer Victor Tatin wrote, 'The glory of having obtained the first results is . . . forever lost to France . . .'

Ironically the first flight that gave the French a boost was the work of a Brazilian! On 23 October 1906 the airship aviator, Santos-Dumont left French soil in his aeroplane, No.14-bis, and in full flight covered a distance over 50 metres. In so doing he won the Archdeacon Cup – the first-ever prize for an aeroplane flight. On 12 November Santos became the first in Europe to fly over 100 metres, this gained him yet another prize.

An 'adopted Frenchman' next soothed the bruised national honour. On 23 October 1907 Henry Farman became the first to fly for 150 metres, and two days later became the first to cover 200 metres. The French Aero-Club now held that: 'The first *authentic* experiments in powered aviation have taken place in France.' The Wrights were almost treated as impostors. In the following year Farman notched up some more firsts and scooped up more prize money. He made the first circular kilometre flight on 13 January 1908. On 30 May he made the first flight in Europe with a passenger, and on 6 July made the first European flight lasting over a quarter of an hour. His machine was a modified French Voisin, but Farman himself remained British until 1937. Still it is fair to say that France was his first love.

While the French congratulated themselves, the Wright brothers were quietly working away at new engines and better machines; too quietly for most of the world, hence the widespread doubts about their achievements. Objectively their conduct was odd. From 16 October 1905 until 6

May 1908, they stayed earth-bound; not a single flight of any of their planes was attempted. Add to that the fact that no one was even allowed to see their machines on the ground, and the reasons for mistrust are obvious. But the Wrights justified this later, by the fact that they were trying to sell their invention to either the US Government or to some other major client. They did, eventually, sign a contract with the US Army and a French company, and with that done, the Wrights took to the air once more and Wilbur came to France to quash the doubters. His performances convinced everyone who saw them. He had truly mastered the art of flying. On 18 December 1908 he became the first man to fly an aeroplane at over 300 feet, he reached 360 feet. On 31 December he became the first man to fly for over two hours, he reached two hours twenty minutes.

Back home in the USA Orville Wright showed the same flair and skills when he demonstrated their machine to the US Army observers. Then, on 7 September 1908 came the first fatal crash in powered aviation. Orville took Lieutenant Selfridge up as a passenger; a crack in one of the two propellers set up violent vibrations; the other prop lost its supports; bracing wires became entangled and the plane nose-dived. Selfridge was killed and Orville seriously injured.

Powered aviation's first fatality.
Lieutenent Selfridge (US Army) dies at Fort Meter. Orville Wright survived

Though they never flew in Britain, the Wrights still had a dramatic impact on would-be British aviators. But it was an American, Samuel F. Cody, who became the first man to fly in the British Isles; on 16 October 1908. He used a biplane clearly based on the patterns used by the Wrights, though it was much sturdier. A number of publications have argued that the Englishman, A.V. Roe should have the credit for the first flight, and in his old age Roe himself wanted it believed that he had flown *before* Cody.

This earlier first flight is claimed for June 1908. But this claim is easily refuted by A.V. Roe's own advert issued in 1910. It appears in the 1910 edition of *The Aero Manual*; it advertises Roe's planes and engines and says: 'A.V. Roe was the first Britisher to fly in Britain, his first machine being patented in 1906.' So in the 1900s he merely sought the honour of being the first *Britisher*, not the first man. But even this claim fails to stand up and the honour goes to J.T.C. Moore-Brabazon, whose first flight was in April 1909. Brabazon then went on to win the *Daily Mail* prize for the first circular mile by a Briton using an all-British plane, that was on 30 October 1909.

1909 was a busy year for new air-records; there were first flights in Austria, Canada, Portugal, Sweden, Rumania, Russia and Turkey. The first rotary engine, a *Gnome,* was put into use. The first cinematograph pictures were taken from a plane in April (near Rome). And the first attempt was made to fly the English Channel. This attempt, by Hubert Latham, was started from Sangatte, near Calais on 19 July 1909. His monoplane, *Antoinette,* developed engine trouble and after covering 7 miles, he had to ditch in the water and float around until rescued. But he was undeterred, and with a new aircraft he returned to Sangatte for a second attempt. At the same time Louis Blériot decided that the first Channel crossing would be his. At 4.35 a.m. on the morning of 25 July, Blériot took to the air while Latham still lay sleeping. The plane, made to Blériot's own design, flew from Baraques (near Calais) heading for Dover. It was deflected off course by side winds and reached Deal, but Blériot stuck to his plan and turned westward towards Dover, where he landed rather heavily but safely. This historic first flight had taken just 37 minutes.

In August 1909 the world's first air-show was staged at Reims. Now the aeroplane was taken seriously for the first time. Its future promises were now fully appreciated. They were bright indeed and not even the first death of a pilot, (Eugene Lefebvre on 7 September) could overshadow them. Indeed the day after Lefebvre's death, Sam Cody made the first flight of over an hour in Britain. Then on 30 December came the first flight of over 100 miles. This lasted over two hours (another first) and was accomplished by Delagrange in a *Blériot XI.*

The pace began to grow hectic in 1910. Over twenty aviation rallies were held in Europe and first flights took place in Argentina, Brazil, China, Indo-China, Spain and Switzerland. On 8 March, Baroness de Laroche became the first woman to qualify as a pilot. The first night flights took place; in the world on 10 March (Aubrun flying a Blériot in the Argentine); in Europe on 28 April (Grahame-White using a Farman). On 2 June, the Honorable C.S. Rolls made the first double-crossing of the English Channel, but did not bother to set down in France; a few cheery messages were thrown down over the buildings owned by the Channel Tunnel Company (yes, they had one then!); then it was back to Dover. Six weeks later on 12 July Rolls was dead, the first British pilot to be killed when his plane broke up and crashed at Bournemouth.

The first passenger was carried across the Channel by John Moisant using a two-seater Blériot, on 17 August 1910. Then two-way ground-to-air wireless communication was employed on 27 August by James McCurdy while circling in the Sheepshead Bay area of New York State in his Curtiss plane. Another Curtiss became the first plane to take off from a ship on 14 November, this was soon capped by the same pilot (Eugene Ely) when he both landed on and took off from the cruiser *Pennsylvania* on 18 January 1911. This was the first such event, and since the cruiser was fitted with a flight-deck and arrestor-ropes, this makes this ship the first aeroplane carrier. Yet another Curtiss aircraft, flown by Glen Curtiss himself, was the first to land on water, taxi and take off again. This was staged on 26 January to convince the Navy and sure enough, this 'hydroaeroplane', as he termed it, became the first aeroplane to be bought by the US Navy. As the Navy A-1 it first flew on 1 July 1911.

Other navies began to explore the possibilities of fleet air arms, and the first torpedo drop was made by the Italians using a Farman biplane piloted by Capitano Guidoni. 1911 also witnessed the development of the first bomb-sight (in the USA) and the first fitting of aircraft with guns. The first actual use of the aeroplane in a real war occurred when a reconnaissance Blériot from Tripoli kept watch on Turkish troop movements at Azizia on 22 October 1911. Just ten days later, on 1 November, another Italian plane dropped the first bombs used in war. Released from an Air Flotilla plane, they landed on Turkish positions at the Taguira Oasis and at Ain Zara. The pilot was Second Lieutenant Giulio Gavotti.

The peaceful possibilities of the aeroplane were well demonstrated by Louis Breguet when he carried eleven passengers in a huge aircraft of his own design. This 3-mile trip made on 23 March 1911, was the first demonstration of airline travel and the first transport of more than two people.

This was soon followed by the first non-stop flight from Paris to London on 12 April 1911. No passengers this time, just pilot Pierre Prier, but it defined the shape of things to come. Another important use for the aeroplane was shown when, on 18 February 1911 a Humber biplane piloted by Henri Pequet carried 6,500 letters from Allahabad to Naini Junction. This was the world's first official airmail flight.

The first regular airmail service began four days later and all its envelopes bear the frank 'First Aerial Post, UP Exhibition, Allahabad, 1911'. In Britain the first official airmail service was timed to commemorate the Coronation of King George V. The first bags of mail were taken up by Gustav Hamel in his Blériot machine on 9 September 1911 and the mail was conveyed from Hendon to the Royal Farm at Windsor. As a service it ended on 26 September. Three days earlier, on 23 September, another Blériot monoplane transported the first official airmail in the USA. It covered a 6-mile journey, starting from Nassau Boulevard, New York. The pilot, Earl Ovington, was sworn in by Postmaster-General Hitchcock and given the official title 'Air Mail Pilot No. 1'. Italy also set up its first airmail service in September. On the 19th, mail began to be flown from Bologna, Venice and Rimini. In truth though, none of these services amounted to much but they did demonstrate the future value of fast goods transport of all types.

Record-breaking, especially for long distances became an ever stronger lure, but of more importance were new methods of aircraft construction. First among these was the introduction of monocoque fuselages; that is, bodies made with skins or shells which carry all or part of the load. Before this innovation bodies were mere frameworks, sometimes open, sometimes covered. The monocoque principles were first applied to a wooden structure by the Swiss designer Ruchonnet then adapted by Béchereau to suit a streamlined plane named *Monocoque Deperdussin*. This was a futuristic design of great elegance; a design destined to break records and carry off glittering prizes. In February 1912 this plane became the first to fly at over 100mph; taken to Chicago it won the Gordon Bennett Cup with a world record of 108.18mph. The pilot, the first man to travel at 100mph, was Frenchman J. Védrines.

Other notable firsts of 1912 include the first enclosed cabin aeroplanes, both monoplane and biplane and both by Avro, and on 1 March, the first parachute drop from an aeroplane, made by Captain Albert Berry over Jefferson Barracks, St Louis, Missouri; then on 16 April Miss Harriet Quimby became the first woman to fly the Channel; and 13 May saw the founding of the Royal Flying Corps, the first step towards an independent British airforce.

1913 witnessed a growing awareness of the aeroplane as a war machine. At Farnborough, the British created the first aeroplane meant for warfare, and nothing else. This was the F.E.2a fighter, a biplane with a pusher engine mounted behind the pilot. It was the pattern for the F.E.2b and the Vickers *Gunbus* fighters of the approaching war. The Russians for their part developed the largest aeroplane ever seen; the first four-engined plane in history. This menacing cabin biplane, with its 92-foot wingspan looked every inch like a giant bomber. But the *Bolshoi* was planned by its designer, Sikorsky, to serve only as a benevolent civilian transport, as an airliner. It first flew on 13 May 1913 and was viewed as a flying hotel complete with a sofa, armchairs and liquid refreshments, probably vodka.

Yet the first airline never offered anything quite as spectacular as a four-engined giant. It operated with a Benoist flying-boat which ferried passengers one at a time across the 22-mile width of Tampa Bay, Florida. This St Petersburg–Tampa Airboat Line opened on 1 January 1914. Pilot Tony Jannus charged $5 a trip and saved his clients the chore of having to make a 36-mile road journey around the bay. No meals were served.

Fittingly the first full meal served in the air was eaten aboard one of Igor Sikorsky's giant four-engined cabin machines, the *Ilia Mouriametz* in June 1914. Two months later the Great War began and civilian pleasures were subordinated to grim military needs. The first British plane of the war to fly over the Channel to aid the French was a BE2a piloted (on 13 August 1914) by Lieutenant H.D. Harvey-Kelly. Others followed and within ten days the first RFC aeroplane was shot down when, on 22 August, German rifle fire disabled the Avro 504 flown by Lieutenant V. Waterfall. Three days later the RFC exacted revenge when Lieutenant Harvey-Kelly and two other pilots used unarmed spotter planes to force a German two-seater aircraft to the ground.

The Russians used more drastic methods to gain their first kill. On 26 August 1914 Staff Captain Petr Nesterov, of the Imperial Russian Air Service, rammed the Austrian plane piloted by Baron von Rosenthal. Both pilots died, making Nesterov the first Russian battle casualty in air warfare.

For a while, aeroplanes were mostly used for reconnaissance and their first crucial use in this role came at the battle of Tannenberg in August 1914. Good German intelligence brought in by spotter planes resulted in a savage defeat of the Russian forces. Hindenburg later wrote '. . . without airmen there would have been no victory at Tannenberg'.

The aggressive phase began on 30 August 1914 when a German Taube dropped five bombs on Paris, the first time a capital city was hit by an aeroplane. This was soon followed by the first missiles dropped by RFC planes, when No. 3 Squadron unloaded showers of 5-inch-long steel darts (called flechèttes) over enemy ground positions. Their impact was more on the mind than on the body. Distinctive war markings were quickly brought in; the famous British roundel was first adopted on 11 December 1914; the equally famous German Iron Cross insignia was first painted on in September 1914.

How the French saw their first victory in the air.
The fight of 5 October 1914

On 5 October 1914 a French Voisin two-seater pusher plane became the first in the world to shoot down another plane. Its machine gun destroyed a German Aviatik over Jochery, near Reims. The first four-engined bombers were, logically enough, the Sikorsky biplanes of the *Ilia Mouriametz* pattern. Their menacing looks became translated into menacing actions on 10 December 1914; though their first bombing raid was not staged until 15 February 1915. The first bombing raid on Britain took place on 21 December 1914, but the two bombs hit the sea off Dover. The first bomb to hit British soil exploded near Dover Castle on 24 December 1914. At sea the need for a British aircraft carrier was met by converting HMS *Ark Royal* while still under construction. She was launched in 1914 as the first purpose-built carrier in the world. Her craft were all seaplanes.

Up until 1915 all fighter aeroplanes were hampered by their propellers. A wrongly aimed burst of fire could rip the propeller to pieces. Guns mounted way out on the wings were a partial solution only, since a jammed gun was a nightmare to clear. On 1 April 1915 French pilot Roland Garros made the first kill with a machine gun that fired *through* the propeller disk. The solution to the problem lay in fitting metal deflector plates to the roots of his Saulnier propeller. Most bullets went through without contact, while those that hit the prop were thrown to one side without causing fractured blades. The secret of his through-the-prop gun became known to the Germans when Garros had to make an emergency landing behind German lines. Once seen, the idea spurred the Germans on to develop the first successful synchronizing mechanism for airborne machine guns. They made *the propeller* fire the gun, but only when its blades were out of line with the stream of bullets. This was one of the most important of all warplane inventions, and it was first used in July 1915 by Fokker E series monoplanes. Its first victory was achieved by Lieutenant Kurt Wintgens using the slightly different Fokker M5K. On 11 July 1915 he shot down a French Morane-Saulnier fighter, the first of many to fall to the 'Fokker Scourge', for synchro-guns gave the Germans air supremacy on the Western Front for all of nine months.

On the Eastern Front the Naval Air Arm pioneered the use of the airborne torpedo. On 12 August 1915 a Short 184 seaplane took off from HMS *Ben-My-Chree*, headed for a lame Turkish supply ship, wallowing in the Sea of Marmara, and launched a torpedo at it. This first use of the torpedo in war seemed to sink this ship, but later on, the captain of British submarine E14 claimed that he had fired a torpedo simultaneously. However, the same pilot, Flight Commander C.H. Edmonds, struck again on 17 August and this time he sank a Turkish steamer just north of the Dardanelles. This is the first undisputed sinking by air-launched torpedo.

The US Navy, meanwhile, though neutral, was honing up its fighting techniques. On 5 November 1915 it catapulted the first aeroplane from a ship, the USS *North Carolina*, while she lay at anchor in Pensacola Bay, Florida. Next day the same plane, an AB-2 flying-boat, with the same pilot, Lieutenant Commander H. Mustin, was catapulted from the ship while it was in motion. The planes, like all the other craft at that time, were of mixed construction: wood, metal and canvas.

The first all metal monoplane to fly took to the air on 12 December 1915. This was the Junkers J1, intended as a close-support aircraft. It was never put into production though, yet it served as a model for further valuable developments. But those developments were far ahead and the air war continued to be fought by mixed-construction craft. This included the first triplane fighter to enter service, the Sopwith Triplane. This British design flew for the first time on 28 May 1916 and it was to inspire the Germans to commission their own triplane, the Fokker F1, which first flew on 30 August 1917.

At sea the US Navy set up the United States Naval Flying Corps on 29 August 1916. They were observing the changing face of naval warfare and preparing for change. Just weeks later came the first fruits of change — the sinking of a submarine by an aeroplane. On 15 September an Austrian Lohner flying-boat bombed and destroyed the French submarine *Foucault*.

It was not until 20 May 1917 that the first German submarine met a similar end. Then, U-36 went to its death in the North Sea after being attacked by an American flying-boat piloted by Flight Sub-Lieutenant C.R. Morrish.

On 28 November 1916 the aeroplane brought war to civilian London when a German LVG CII dropped six bombs over the capital. These, the first from a plane, landed near Victoria Station. But this was a pinprick compared with the first mass bombing raid using twenty-one Gotha heavy bombers. They struck at Folkestone, Shorncliffe and other towns on 13 June 1917, killing ninety-five people and injuring some 260 more. London's first mass bombing raid came on 13 June 1917; seventy-two bombs hit the capital leaving 162 dead and 432 injured, making it the worst bombing raid of the war. On 2 September 1917 the Germans changed tactics and made the first night-time mass bombing raid on Britain, hitting Dover and its surrounds.

1918 saw the formation of Britain's first fully independent airforce when, on 1 April, the RFC and the Royal Naval Air Service combined to form the Royal Air Force. In France the Americans first arrived with their 95th Aero Squadron on 18 February; they flew their first patrols on 19 March; made their first kills on 14 April (two German planes), and made their first bombing raid on 12 June. On 11 September General William Mitchell of the US Air Service took command of the first giant aeroplane operation of the war when a mixed force of 1,483 aircraft was employed during the battle for the St-Mihiel salient.

Two months later the First World War ended. Officially the first minute of peace began at one minute past eleven o'clock (p.m.) on 11 November 1918, and for most of the world that is correct, but in Africa the war did not end until fourteen days later, when Paul von Lettow-Vorbeck surrendered his forces. So the first day of peace in Africa stands at 25 November 1918.

With peace came the slow development of civil aviation. The first mail and passenger services between Paris and London were started on 10 January 1919, but these were limited services intended only for the smooth running of the Versailles Peace Conference. The first passenger service open to the public was inaugurated on 5 February 1919 by the Deutsche Luft-Reederei of Berlin. It ran daily services between Berlin and Weimar, a distance of some 120 miles. In Britain the first civil airline began services on 10 May 1919. It was a short-distance service only, between Manchester, Blackpool and Southport, run by A.V. Roe & Company, using its Avro biplanes. The first long-distance services began on 26 May 1919 and were organized by the North Sea Aerial Navigation Co. This enterprise came to make the first regular air links between Hull, Leeds and London (Hounslow). Hounslow was also the base for the world's first daily international flights. Organized by Aircraft Transport & Travel Ltd, they used de Havilland DH16 planes to shuttle four passengers at a time between Hounslow and Le Bourget, Paris. They began on 25 August 1919.

Intercontinental services were first operated by the French Lignes Aériennes Latécoère on 1 September 1919. They flew Breguet 14s craft between Toulouse and Barcelona, then overseas to Tangier. Oddly enough the USA was late in organizing its civil aviation industry. It had such efficient rail, road and waterway systems that the need did not seem so pressing, thus the main interest lay in using aircraft as mail carriers. The first stage of the US transcontinental airmail network was opened on 15 May 1919. This short stage, between Chicago and Cleveland was opened to attract attention to the larger plan of a coast-to-coast airmail service at standard first class rates. (That service first began from San Francisco on 22 February 1921.)

The USA did clock up one first of special note when an American Navy Curtiss NC-4 flying-boat crossed the Atlantic (8 May–31 May). But it did this in a series of thirteen hops from one port to another, and some 200 miles of the trip were taxied on the surface of the water, rather than covered in the air.

Of far greater significance was the first non-stop crossing of the Atlantic on 14–15 June 1919. Using a Vickers Vimy twin-engined bomber, Captain John Alcock and Lieutenant Arthur Whitten Brown set off from St John's, Newfoundland and landed at Clifden, County Galway in Ireland. The flight took sixteen hours twenty-seven minutes. Both men were knighted, but sadly Sir John Alcock was dead before the year ended, killed in a flying accident on 18 December.

The reliable Vickers Vimy was next used to make the first flight from Britain to Australia. It took from 12 November to 10 December 1919 to complete the trip from Hounslow to Darwin. Appropriately the four-man crew was headed by two Australian brothers, Captain Ross Smith and Lieutenant Keith Smith. They were knighted for their feat and awarded a prize of £10,000. Yet another Vickers Vimy was employed on

The US Navy Curtiss plane was first to reach 250 mph.
It was piloted by Lieutenant Brown, September 1923

the first ever flight from Britain to South Africa, which started on 4 February 1920. This time the 'old reliable' developed trouble over Wadi Halfa, in the Sudan and crash landed. After an eleven-day wait a second Vimy (courtesy of the South African Government) was ready and the flight continued only to end in a second crash at Bulawayo in Southern Rhodesia. The toil restarted on 17 March, this time in a DH9, and on 20 March the goal was reached at Wynberg Aerodrome, Cape Town. The intrepid aviators, Lieutenant Colonel Pierre van Ryneveld and Squadron Leader Chris Brand, were given knighthoods and £500 prize money.

From a strictly technical viewpoint, some of the far-reaching developments of this period include: the first automatic pilot installed in a British commercial plane, the Aveline Stabiliser of 1920; the first practical retractable undercarriage, fitted to the Dayton-Wright RB Racer of 1920; the first pressurized cabin, used on the de Havilland D.H.4 in the US (1922); the first variable-pitch propeller, the Turnbull prop (USA) of 1923; and the first use of grounded electric beacons to direct night-flying planes. These five innovations were to become standard requirements in later years. Also of great importance was the first successful in-flight refuelling exercises which began on 27 June 1923, at San Diego, California. Fifteen in-flight refuellings enabled Smith and Richter to set up a world endurance record of thirty-seven hours fifteen minutes, using a de Havilland DH4B. This in-flight technique is now a priceless asset, especially in combat, as the Gulf War proved repeatedly.

On the strictly peaceful side, national airlines were first set up in the Soviet Union in March 1923 (Dobrolet); in Czechoslovakia on 1 March 1923 (CSA); and in Britain on 1 April 1924 (Imperial Airways). The first round-the-world flight on 6 April to 28 September 1924, showed the vast potential of air travel, even though the flight was conducted by military Douglas DWCs. Then a further use was found for the aeroplane. It became an unexpected friend of the farmer when the first aerial crop-dusting firm, Huff Daland Dusters, set up in 1924, at Macon, Georgia USA.

As for air travel, there was no dramatic upsurge of interest, just a slow increase in passenger numbers and routes on offer. Even the first flight over the North Pole on 9 May 1926 (by Byrd and Bennett of the US) was just a passing side-show. But the first solo crossing of the Atlantic on 20–21 May brought about a great mood of excitement and a massive popular interest in air travel. Overnight the pilot, Charles Lindbergh, became a world hero. His thirty-three-hour flight in the single-engined Ryan *Spirit of St Louis* earned him entry to Buckingham Palace; and President Coolidge sent a US warship to bring him home. And there the President gave him the DFC, while Congress struck a special medal for him. 'Lucky Lindy' then became the flying ambassador for civil aviation. He flew his plane from one American venue to another. Whole towns stopped work to see and cheer, and this seething interest was so great that by 1928 the number of US civil air companies doubled, from twelve to twenty-five. Amalgamations followed and by 1929, the US was first in the world, carrying more of everything by air than everyone else put together.

At this time people began to take note of an aviation novelty – the autogyro. This was Juan de la Cierva's brainchild. He saw horizontal revolving blades as the secret of short take-off runs and landings. They also gave a plane the ability to fly very slowly without stalling. His first machine, a C4, flew successfully on 9 January 1923. His first two-seater flew on 29 July 1927 and the next day it flew with a passenger for the first time. His improved C8L then became the first rotating wing-craft to fly the English Channel on 18 September 1928. Juan de la Cierva himself was the pilot and with him he took a passenger. The autogyro differed from the helicopter because its rotating blades were not driven by the engine to provide lift. Its engine simply provided forward drive; air pressure spun the vanes. It was not until 1936 that the problems involved in helicopter design were solved. First to build and fly a successful helicopter was Louis Breguet in early 1936. Though it flew for an hour without mishap, it was never taken any further. The first helicopter to succeed and influence future production was the German Focke-Wulf FW 61 twin-rotor machine. Its first prototype flew on 26 June 1936 and led to the first government contract for helicopters. But Focke-Wulf were expected to develop larger machines able to carry useful loads. The FW Drache, capable of carrying twelve people, resulted from this contract.

The 'Autogiro' invented by Señor Juan de la Cierva

The 1930s were the heyday of the flying idols. Amy Johnson, Amelia Earhart, Wiley Post, all joined Lindbergh in sharing the popular adoration. Johnson was hailed for her feat in becoming the first woman to fly from Great Britain to Australia (5 to 24 May 1930). Earhart was acclaimed for her Atlantic solo flight, the first for a woman, on 20–21 May 1932. Wiley Post was applauded for his first solo flight around the world, between 15 and 22 July 1933. Sadly all three were later to perish in crashes.

Many of the problems faced by the record breakers, bad weather conditions, icing up of flaps, navigational drift and so on, led designers to create sound remedies. Automatic pilots were developed up to the point where, on 23 August 1937, a plane was landed at Wright Field, Ohio, on automatics alone. Pressurized cabins were finally made fully efficient by Lockheed, in 1937, and on 7 May 1937, a Lockheed XC-35 using the new cabin flew for the first time. Then the pressurized cabin was built into the Boeing 307 Stratoliner, the first civilian airliner able to fly above most rotten weather conditions. The 307 prototype first flew on 31 December 1938. As the 307B it was the first to enter service in April 1940, when it became part of the Trans-Continental and Western Air fleet.

Before the 30s ended, the Spanish Civil War had been used as a testing ground by the German, Italian and Russian military machines. The famous *Stuka* dive-bomber was first used in that war in 1937 and proved devastating against troops without fighter aircraft cover. As a ground-attack bomber it still proved a most effective *blitzkrieg* weapon when the Second

World War broke out, but it soon lost its fearfulness when confronted by avenging *Spitfires* and *Hurricanes*. The *Hurricane* had the distinction of being the first British monoplane fighter. First designed in 1934, by Sydney Camm, it made its first flight on 6 November 1935 and first entered RAF service (11th Squadron) in December 1937.

The first British air raid of the War was confined to leaflet dropping by ten *Whitley III* bombers on 3 September 1939. These first leaflets (6,000,000 in all) fell on the Ruhr, Bremen and Hamburg. The first bombs followed a day later when ten Blenheim bombers raided the German Fleet as it lay off Wilhelmshaven. On 20 September 1939 the first German aircraft was shot down by the airgunner (Sergeant Letchford) of a two-seater *Fairy Battle* flown by No. 88 Sqd RAF. The 'kill' was a Messerschmitt Bf 109E. The first raid over Berlin was made on 1 October 1939, when *Whitley* bombers again dropped propaganda leaflets. And it was not until 19 March 1940 that the first bombs were deliberately aimed at targets on German soil. This time *Whitleys* hit the seaplane base at Hörnum.

The first German planes to hit Britain dispensed with leaflets and went straight for high explosives. On 13 November 1939 four German aircraft bombed shipping at the Shetland Islands, but there were no casualties. The first Luftwaffe attacks on the British mainland began on 1 July 1940; these can be considered the opening parries of the Battle of Britain. With the battle well underway, the first bombs landed on London. They were originally not planned to hit the capital, but one of the German bombers missed its allotted target and jettisoned its bombs at random. That was on 24 August 1940. Swift retaliation followed on the next day when twenty-nine Wellingtons dropped the first bombs on Berlin. Hitler then declared wholesale air war against Britain and the first deliberately planned raid on London took place on 7 September 1940.

German bombs first fell on Russian targets on Sunday 22 June 1941, the first day of the opening of war against the Soviet Union. Konovo, Kiev, Sebastopol and other cities were hit. One month after, on 22 July 1941, the Luftwaffe made its first assault on the capital, Moscow. Russia retaliated by hitting the German capital with its long-range bombers on 9 August 1941.

Neutral America was plunged into the war when Japan made its first ruthless air raid on Pearl Harbor on 7 December 1941. On 10 December Japanese aircraft made their first bombing raid on British targets when they hit warships of the British Far East Fleet in the South China Sea. The *Prince of Wales* and the *Repulse* were sunk.

All bombing raids on both sides involved larger and larger numbers of craft and the manufacture of deadlier bombs, culminating in the 22,000lb

Grand Slam bomb, the largest of the war, first dropped on 14 March on the railway viaduct at Bielefeld. But nothing proved as deadly as the revolutionary atomic bomb. A preliminary test in New Mexico on 16 July 1945 had shown that the first atomic bomb worked. The larger airborne A-bomb was then shipped to the Mariana Islands, loaded into the bomb bay of a B-29 flown by Colonel Paul W. Tibbets and dropped over the city of Hiroshima. This first air-dropped atomic bomb killed 71,000 people, severely injured 68,000 more, and razed some 4.7 square miles of the city. The atomic age began with a howl of anguish. But as far as planes were concerned, no one thought in terms of atomic-powered craft. The limits of propeller drives were already being dealt with by the development of jet-propulsion units and exotic rocket-drives.

JET PROPULSION

The idea took shallow root long before the age of flight; indeed the first feasible scheme for jet-propelled transport is usually credited to Sir Isaac Newton. Taking time off from apples and gravity, he forecast that people would one day travel at the rate of 50mph. And many sources claim that in 1687 he even described a steam road-locomotive that would pioneer fast travel. But the claim is erroneous. The steam-carriage proposal first appeared in a book by Gravesande, a popularizer of Newton's ideas. The design set out to demonstrate Newton's third law of motion: 'To every action there is opposed an equal and opposite reaction', and it used the jet propulsion principle. It was an uncomplicated design. A four-wheeled chassis would support a circular boiler and firebox at its middle. The boiler would have a long, tapered steam-exit nozzle pointing backwards. In front, back to the boiler, would sit the driver. A long shaft linked to a

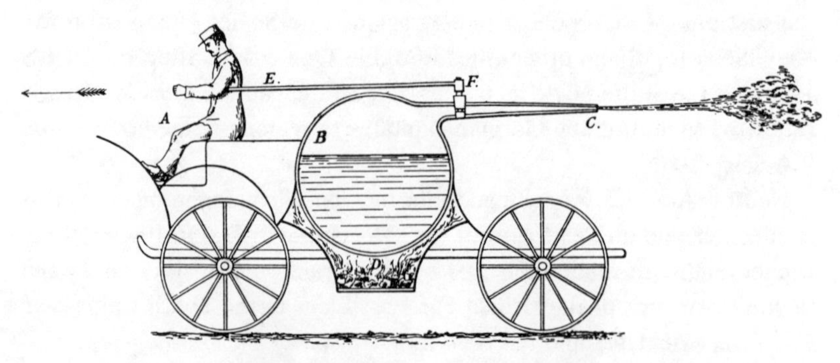

Jet propulsion; the first feasible scheme

valve in the nozzle would give the driver full control of the steam jet, allowing him to stop and start, and vary the speed of travel at will. Mercifully this locomotive never reached the open roads, where its scalding jet would have created havoc. But for all that, the basic plan embodied sound ideas.

The first practical use of Jet Propulsion came in 1786, when James Rumsey succeeded in driving a boat against the current of the Potomac at the rate of 4mph. Rumsey used his steam-engine to operate a pump which forced a jet of water rearwards. The reaction drove the boat forwards with a firmness that overjoyed both the inventor and his distinguished observer, General Washington. Yet the system never caught on. It survived only in small model boats meant for children.

Failure at sea was completely overshadowed by success in the air. The limits of propeller-driven aircraft revived interest in jet propulsion as a prime mover and the 1930s saw the first crucial moves towards jets. The concept had, however, been first used in aircraft design as long ago as 1865 by French engineer Charles de Louvrie. The design for his *Aéronave* involved the burning of a vaporized petroleum oil, ejected through two rear jet pipes. And the first jet-plane built (though not flown) dates back to 1910 when H. Coanda showed his biplane at the Paris Salon. It was equipped with a reaction propulsion unit based on a 50hp Glerget engine driving a large ducted fan.

Twenty years later in 1930, Frank Whittle placed jet research on a sound basis with his first patent, applying the gas turbine to power a jet-propulsion engine. This was the first use of the gas turbine in this context.

On 12 April 1937 Flight Lieutenant Whittle demonstrated the world's first turbojet aircraft engine, a B.T.H unit mounted on a test bench. Following this, the first Air Ministry contract for a Whittle engine was issued, in March 1938. In 1939 the first jet-propelled aircraft were ordered from the Gloster Aircraft Company, and on 15 May 1941, the first successful jet flight was made in Britain. This success led to the export of a Whittle engine to the USA, where it was made under licence and fitted to the Bell XP-59A. The Bell jet, America's first, first flew on 1 October 1942.

Though Britain was first in the field in planning for jets, the credit for staging the first-ever flight of a jet-powered aircraft must go to Germany. In June 1938 the Heinkel HeS 3B turbojet engine, designed by Doctor Pabst von Ohain, was used to power a Heinkel HE 118 specially modified as a flying test-bed. It was not though, a production prototype but a utilitarian alliance between a radical engine and a conventional airframe. Even the later gas-turbine jet plane, the He178, was little more than an experimental craft of conventional design. Its first display flight was on 27 August 1939, when it was piloted by Flugkapitan Erich Warsitz. An earlier, shorter and secret flight is said to have taken place three days earlier.

The first flight of an aircraft designed only with jet-flight in mind took place on 27 August 1940. The plane was the Italian Caproni-Campini CC2, which made a ten-minute excursion from its base at Forlanini aerodrome. On 30 November 1941 the CC2 then made the first jet-flight of realistic proportions when it flew from Milan to Pisa, then on to Rome. Like the Coanda machine it used a piston engine driving a ducted fan, and that proved its great weakness. As experience was soon to show, full efficiency lay with turbine-driven jets, so this revolutionary plane was put to rest and never saw combat.

First purpose-built jet aircraft – Italy's Caproni-Campini CC2

The world's first jet combat aircraft was the Heinkel He280. This first flew on 2 April 1941, but all its air fights were destined to be mock-battles, staged in 1942. It was dropped in favour of the rival Messerschmitt project. The He280 was also the first twin-engined jet aircraft.

The first jet-engined fighter and fighter-bomber to be used in action was the Messerschmitt Me-262, a twin-engined, sleek plane with a top speed of 541mph. It first flew on 18 July 1942, but it saw no action until July 1944, by then far too late to have any real impact on the war in the air. Though a formidable threat to most conventional prop-aircraft, it was successfully combated by the prop-driven Hawker Tempest Mk5.

In Britain the first combat jet was the twin-engined *Gloster Meteor*, first flown on 5 March 1943. The first RAF unit to use the planes was the 616 Squadron, then based at Manston, Kent. 616 Squadron first used its *Meteors* on combat flights on 27 July 1944. The first kill recorded by this plane was on 4 August 1944, when a V-1 flying-bomb was intercepted. After the end of the war the Meteor F.1 was re-engined with two turboprop units driving two five-bladed propellers of just under 8 feet in diameter. On 20 September 1945 it made the first ever flight of an aircraft fully dependent on turboprops.

In its pure jet form the *Meteor* set the first post-war absolute world speed records. On 7 November 1945, *Meteor* EE454 reached the record speed of 606 mph. This was exceeded on 1 September 1946 by *Meteor* EE549 which reached 616 mph.

Since the Meteor was judged unsuitable for carrier work, the first jet aircraft to operate from an aircraft carrier became the smaller, single-engined De Havilland Vampire 1. Landings and take-offs, from the deck of HMS *Ocean*, began on 3 December 1945. Seven months later the first American carrier landings and take-offs took place on the USS *Franklin D Roosevelt*. The jet fighter used was the McDonnell FH-1 Phantom; the date was 21 July 1946.

The first jet plane to shoot down a jet-propelled opponent was a Lockheed F-80 piloted by Lieutenant John Russell of the US Air Force. On 8 November 1950 he engaged and shot down a Soviet MiG-15 over North Korea. The Lockheed F-80 *Shooting Star* was also the first turbojet-powered fighter to enter service with the US air forces. It first flew on 8 January 1944, but was too late to join in the Second World War. Its triumphs were reserved for the Korean conflict.

The new breed of jet fighters ate up fuel at a rapid rate, hardly fitting them for long-distance records. But on 22 September 1950 a USAF Republic EF-84 Thunderjet fighter made a non-stop crossing of the North Atlantic. It gained this record by refuelling three times while in flight, using the hose-and-drougue system linked to a tanker plane.

The first jet aircraft to make a non-stop, North Atlantic crossing without refuelling, was an English Electric *Canberra* B.Mk-2 bomber. It flew from Britain to Baltimore USA on 21 February 1951. The *Canberra* was also the first jet bomber to be manufactured in Britain, the first to serve with the RAF, and the first non-American design to be used by the post-war USAF.

Jet propulsion led to new attitudes towards wing design and the Americans tried to use aircraft with variable-geometry wings, allowing the pilot to alter wing-span and configuration to suit different altitudes. The first plane to use such wings was the Bell X-5 which first flew on 20 June 1951. Others regarded the move as too risky and opted for fixed, delta-wing designs. In Britain the Gloster Javelin was developed as the first twin-jet delta fighter in the world. It first flew on 26 November 1951.

The 'jet-setters' came into being when BOAC inaugurated the first passenger service in the world dependent on jet-aircraft. They used the De Havilland DH Comet H1 on their London–Johannesburg route beginning on 2 May 1952. This was quickly extended to their London–Ceylon and London–Singapore flights. Flying times were almost halved. There were temporary set-backs after two mysterious crashes, but investigations showed that metal fatigue caused by the pressurization of cabins was the cause. New standards dealt with that problem and in 1954 the first US Boeing 707 jet liner made its first test flight. The 707 began its passenger service in 1958 as the first truly intercontinental airplane. Ten years later Boeing introduced its first *Jumbo Jet*, the 747, the largest commercial aircraft in the world and the first in importance for intercontinental transport of goods.

Once the jet plane had shown its superiority in so many ways it became clear that the logical next step would involve passenger planes able to fly several times faster than the speed of sound. There were busy, or

Boeing's 747, the world's first wide-bodied jet airliner – the 'Jumbo'

impatient, or just idle-rich people willing to pay for such a service. With this in mind the Anglo-French supersonic aircraft scheme was established. It produced the amazing *Concorde,* the first supersonic passenger plane, flying at twice the speed of sound. It was the first of its kind to enter service in 1976. Sadly its only rival for first place, the Soviet *Tupolev* 144, met with disaster at the 1973 aviation 'Salon' in Paris. It took off on a demonstration flight meant to impress the world and, within minutes, exploded in the air and crashed to earth.

Today designers dream of aircraft able to make Concorde look like a slow-coach. And they have in mind craft that look like flying cities. But all dreams have to face the harsh daylight of economics. And economics show us that there is little possibility of the twentieth century seeing the first of these flying cities. So we have to stay content with flying village halls. And why not?

THE ROCKET BREAKTHROUGH

Just as jets overcame the limits of props, so rockets overcame the limits of jets. Only the rocket had the capacity to take Man into the Space Age; but the rockets that conquered Space differed greatly from the rockets first used in the thirteenth century by the Chinese. The rockets of the Middle Ages depended on the burning of solid fuels, exactly on the lines of the rockets used in modern firework displays. They were useful for carrying incendiary and explosive warheads, but they were not used in any force until the nineteenth century. On 8 October 1806 the first use of war-rockets in great numbers took place when the British fleet attacked Boulogne, the assembly point for Napoleon's 'invasion fleet'. Some 2,000 rockets rained down on the town, set it on fire and thwarted the invasion plans. These rockets were developed by Sir William Congreve and first tested in 1805.

Oddly enough the war-rocket is still remembered daily in the USA since it features in their National Anthem ('the rockets red glare'). This makes *Old Glory* the first, and only anthem to mention the rocket.

The first sober writings on modern rocketry were the work of an obscure Russian schoolteacher, Konstantin Eduardovich Ziolkovsky. Years of calculations and deductions on the future of space travel were finalized in 1898, but not published until 1903. The first man to translate ideas into actions was the American, Doctor Robert H. Goddard, Professor of Physics at Clark University, Worcester, Massachusetts. It has been asserted that it's impossible to design, launch or construct a rocket today without infringing one or more of his 214 patents! By 1925 Goddard was the first to develop a rocket motor using liquid propellants.

He used petroleum spirit and oxygen. On 16 March 1926 he became the first to design, construct and launch successfully a liquid-fuel rocket. On 17 July 1929 he sent the first rocket-borne cargo aloft; a barometer and thermometer with a monitoring recording camera.

Prior to his practical demonstrations, Goddard had received the first US patent on multi-stage rockets (1914); had published the first basic mathematical theory underlying rocket flight and propulsion; and had proved experimentally that a rocket would provide thrust in a vacuum. Other Goddard firsts include: self-cooling rocket motors; variable-thrust rocket motors; practical rocket-landing devices; and rocket fuel-pumps. While on 9 November 1918 he presented the US Army with the plans for the first hand-held rocket weapon, the *Bazooka*. These Bazooka plans were shelved until World War II, when they were put to use with spectacular effect.

The liquid-propelled rocket motor first showed its transport possibilities in Germany in 1928, when Max Valier drove his rocket-powered racing car. In the same year, in the same country, on 11 June, Friedrich Stamer first flew a mile in a rocket-driven Lippisch glider. In Germany again, in 1936, Wernher von Braun first mounted his liquid-propellant engine on a small Junkers aircraft. Though it was ground-tested only, this was a vital step in the designing of the first *Vengeance Weapon* — the V-2.

The precursor of the V-2, named the A-2 rocket, was first launched in December 1934 and reached an altitude of 1.5 miles. But many problems had to be overcome before the first V-2 rocket was launched on 6 September 1944. In the meantime the first specially designed rocket-driven aeroplane (the Heinkel He176) was successfully flown on 20 June 1939. Within a year the first liquid-fuel rocket aircraft was ready. As the DFS194, it first flew in August 1940; it was then taken up by the Messerschmitt company. On 13 August 1941 Messerschmitt demonstrated the first valid rocket-powered fighter, the Me163 Komet, in a full-power flight. The Americans were next when they first flew their rocket-powered Northrop MX-324 fighter on 5 July 1944, while the Japanese developed the first purpose-built suicide aircraft, the rocket driven Yokosuka MXY-7, and put it to use on 21 March 1945. With the war over, the Americans concentrated on supersonic flight and on 14 October 1947 their Bell X-1 rocket-plane became the first in the world to achieve manned supersonic flight, when it reached 670mph, or Mach 1.015. The pilot was Captain Charles Yeager USAF.

Military rivalries between the USA and the USSR culminated in the rocket-launching of the first man-made satellite, Sputnik 1, on 4 October 1957. The next stage in space exploration saw the launching of the first living creature into orbital flight, the Russian bitch 'Laika' —

on 3 November 1957. And the race began to place the first man in Space.

The Americans lagged behind; their first satellite, *Explorer*, was not launched until 31 January 1958. But their satellite, US *Discoverer 6*, became the first to take television pictures of the Earth on 7 August 1959. This feat was soon eclipsed when the Russian satellite *Luna 2* made the first impact-landing on the Moon, on 14 September 1959. And by 19 August 1960 the Russians were able to send two live dogs into orbit and bring them back to earth alive. Next went the dog's best friend – Man, when Russian cosmonaut Flight Major Yuri Gagarin completed a single orbit of the Earth on 12 April 1961. His craft was the Soviet *Vostok 1*. The USA swiftly followed by sending Alan B. Shepard into space on 5 May 1961. By 3 February 1966 the Russians were able to send their *Luna 9* space probe to make the first soft landing on the Moon. The US landed its first Moon probe four months later on 2 June 1966. In 1967 a second US probe landed and made the first analyses of the Lunar soil.

But the most spectacular first of all time came when men first walked on the Moon on 20 July 1969. Using a rocket-powered Lunar Exploration Module, three US astronauts were able to leave the spaceship *Apollo 11* and touch down on the Moon's surface. Then Neil Armstrong and Edwin Aldrin stepped out of their LEM and joyously bounced around in the Moon-dust. Millions watched the breathtaking sight on television, making this the first live broadcast performance from any soil outside of Earth.

The USSR's great first came in April 1971 when it launched its *Salyut*, a manned orbital space-station. Four months earlier the Russians had landed the first probe on the surface of Venus, but it took another four years before further *Verena* probes were able to send back the first images from Venus, this was on 25 October 1975.

For its part the USA sent flights over Mercury (March 1974); over Jupiter (December 1973) and Saturn (December 1974). Then on 20 August 1977 it sent the probe *Voyager 2* to pass near Jupiter and Saturn and provide the first close-up photographs of Uranus.

Until 1981 all rocketry meant for space exploration was expendable. It either burnt out on return, or lost itself in the outer vastness. Then on 12 April 1981 the US launched the first recoverable and reusable rocket-ship, the NASA space-shuttle *Columbia*. Since that date we have grown so used to the idea of space travel and shuttle launches, that the early buzz of excitement has vanished. But, although it *seems* that we are destined to remain that way, take heart – when the first man sets foot on Mars it's sure and certain that the old-style fervour will return overnight.

• PHOTOGRAPHY •

Before the camera, with its ability to record pictures, came the camera obscura, a tantalizing invention whose images were fleeting. The first published illustration of a camera obscura dates from 1545, though the principles were known to Aristotle. In brief, it was a device which allowed rays of light to pass through a small hole and cast an image on a wall or sheet. More complex models used a bi-convex lens in place of the hole, but in all cases the viewed image was upside down. In 1636 Professor Daniel Schwenter of Altdorf University used a complex unit of two lenses of different focal lengths mounted in a tube drilled through a hollow wooden ball. The ball could revolve and cover all directions. When fitted to a window shutter or door this could give a whole series of views from outside. In this form many artists found a use for it in drawing panoramic views. And to free themselves from the restrictions of rooms, they then came to use portable camera obscuras made in the form of bell tents. Next came even more portable forms made as wooden boxes using the reflex principle; a mirror redirected the rays on to a screen of oiled paper (Johann Sturm, 1676). This type was further refined in 1685 when Johann Zahn replaced the paper with an opal-glass screen and painted the interior of the box and lens tube with a matt black. All these refinements were to be used when the photo-camera came into being.

By the eighteenth century the camera obscura was well known and made in a great variety of types, even built into the handles of walking sticks. But still the images were transient. Then a discovery by Professor Johann Schultze of the University of Altdorf gave hope that this could all be changed. In 1725 he discovered for the first time that silver salts darkened, due to the action of *light.* Before then it was held that heat or simply the air itself blackened the silver nitrate particles. He used this knowledge to make the first known light-activated picture. It was one that could not survive, but while it lasted it proved his theory. He took a glass flask, filled it with a solution containing chalk and silver nitrate, and covered it with a paper having letters cut out of it, something like a stencil. As the sun's rays passed through the cut-out letters so they darkened the particles held in suspension. The original cut-outs were now seen as dark letters seemingly imprinted on the solution itself. If the flask was shaken the image promptly disintegrated, but the basic chemistry of film coating had been established. And the results were first published in the 1727 transactions of the Imperial Academy at Nuremberg.

After that came a search for permanent images. And in England Thomas Wedgwood, son of the renowned potter Josiah Wedgwood, used his knowledge of Schultze's work to mount his own experiments using silver nitrate. He made copies of insect wings and leaves by placing such objects on to sensitized pieces of white leather. His friend Sir Humphry Davy did the same and the results were first published in the June 1802 issue of the *Journal of the Royal Institution*. Their pictures disappeared though, if they were further exposed to light. Neither man knew of a way of fixing the image. In France Nicéphore Niepce tried a very different approach when he turned to a substance which *hardened* when exposed to sunlight. He used bitumen of Judea and coated glass plates with it. On this sensitive ground he captured the first successful photocopy in July 1822, using an oiled engraving placed on the plate and exposed to sunlight. He then used coated pewter plates and exposed them to the images cast by his camera obscura. One of the results of this phase, a view from his window in 1827, still exists and counts as the first known camera photograph. But his technique was not the real answer. His images were poor and the exposures needed were several hours at least.

But his work interested another French image-chaser Louis Jacques Mandé Daguerre. Daguerre worked with Niepce until 1833, when Niepce died. Then he continued his research on his own and triumphed by creating direct positive pictures on silver-coated copper plates. He used silver iodide as his light-sensitive coating, and mercury vapour as a developer. His first efforts were shown privately to distinguished artists and scientists late in 1838. The first public disclosure of his methods came on 7 January 1839 before the meeting of the Académie des Sciences in Paris. His Daguerreotype method was beyond dispute the first practicable photographic process, and it gave satisfying pictures of great detail. But it lacked the essentials for a system able to expand. All its marvellous pictures were single offerings. There was no way of duplicating them time after time, since no negative existed. This defect was remedied by the invention of photography as we now know it: the negative-positive process.

At St-Loup-de-Varennes is a stone monument inscribed with the words: 'DANS CE VILLAGE NICÉPHORE NIEPCE INVENTA LA PHOTOGRAPHIE EN 1822'. But we know that to be wishful thinking, based on misunderstanding. Yet who for certain deserves the honour? There are a number of conflicting claims for the status of being the first to invent photography. The Swiss have nominated Friedrich Gerber, a professor at Berne University, for this honour. As early as 1836 (so he wrote in 1839) he was able to fix camera obscura images on paper coated with silver salts. Unfortunately no one seems to have seen examples of this work. The only photographic images ever shown by Gerber were of objects laid on paper. But Wedgwood and others had already done this.

Another claimant is Frenchman Hippolyte Bayard. In 1839 he made a series of photographs in Paris using a direct positive method. He first showed these at an exhibition in June 1839. It is said that Daguerre paid him 600 francs to delay publication of his method. As a consequence, Bayard had to wait until 24 February 1840 before making his public disclosure, and by that time everyone was talking about Daguerre's achievements, so Bayard was completely eclipsed.

The French priorities are contested in Germany, where two scientists, Franz von Kobell and Carl August von Steinheil, are credited with the first photographs. They are said to have reached this stage in 1837, but there is contrary evidence that their involvement did not begin until March 1839, while their first paper negatives were not even shown until 13 April 1839, by which time the English nominee, William Fox Talbot, had not only been photographing for years, but had actually published a paper on his work on 31 January 1839. Fox Talbot's work began in the spring of 1834 and one of his first, small wooden cameras took the first pictures in the summer of 1835. At first he was content with negative images, which he fixed with a strong salt solution. Then he hit on the idea of using a paper negative to create a positive. His simple solution involved pressing a fixed negative on to a sensitized paper, clamping this between glass and exposing the sandwich to sunlight. This was the first

Fox Talbot's first experiments. His 'photogenic drawings'

use of the vital negative system. Fox Talbot did little to advance from this stage until he learned of Daguerre's work, and this spurred him into fresh action. On 25 January 1839 Michael Faraday showed examples of Talbot's work to the Royal Society. This first photographic exhibition was followed on 31 January by the first public lecture on his methods by Fox Talbot himself. In this lecture he gave the first information about the use of photography in recording the images seen through microscopes. This is the first outline of the art which later became photomicrography.

At the time of these lectures Talbot was still calling his images 'photogenic drawings'. It was his admirer Sir John Herschel who first thought of using the name 'photography' and Talbot embraced the new term at once. But the general public first heard of his work under the old name, for within three months of his disclosures the first 'Photogenic Drawing' outfits were on sale. This boxed outfit including chemicals, paper, printing frame and instructions, was made and sold by Ackermann and Company of London. And on 20 April 1839 the reading public saw its first representation of a photogenic drawing in the form of a facsimile of one of Talbot's grouped fern leaves. This appeared on the front page of the magazine *The Mirror*. The same magazine issued the first advertisement for a camera in its June 1839 issue. This was placed by Francis West, an optician from Fleet Street and it offered a 'New Heliographic Camera, with brass adjustments, adapted to Mr Fox Talbot's Photogenic Drawing'. All this was taken rather coolly in England, but in Paris the public was caught up in a frenzy that became known as 'daguerreotypemania'. Daguerre had published full details of his methods on 19 August 1839, and within hours hundreds were queuing at chemists and opticians to buy the essentials for the new hobby. In the USA a copy of Daguerre's manual reached Samuel Morse and his friend D. W. Seager that autumn. A camera was built post haste and the first American photographs were taken. Morse later said he was the first US photographer, but the first US photo exhibited was by Seager. This was of St Paul's Church on Broadway and Fulton Street, New York. Since Morse is known to be unreliable and not too truthful, it is probable that Seager was the first.

So the first photographic parlours were dominated by the French method, but on 21 September 1840 Talbot made a great discovery that reduced exposure times to a few minutes; up until then exposures took twenty or more minutes. His break-through lay in the use of gallic acid added to the silver nitrate solution. This made visible the invisible or latent image that was laid down very quickly on exposure. This first control of the latent image gave the Talbot system a fair chance in competition with the French method. By August 1841 he had developed his methods to the point where he was able to license Henry Collen to become the first professional photographer. At that time though, he was called a Calotypist, since Talbot's pictures were now known as *Calotypes*. In the USA the Calotype rights were bought by Frederick and William Langenheim who opened the first real photographic studios in mid-1849. Five states of the USA took up the Langenheim's licences, but the more greatly detailed daguerreotypes had a far greater appeal and the Langenheims became bankrupt.

In Britain Fox Talbot had set up his first printing establishment in Reading in 1844. From there he planned the world's first book illustrated with photographs. He called it 'The Pencil of Nature' and issued it in six parts between June 1844 and April 1846. Each volume contained twenty-four calotypes, printed at Reading. And the text showed that Talbot was the first to understand the future role of photography as an art quite distinct from painting or engraving. He viewed it as a recording system, first and foremost. He was the first to suggest using photographic records of valuables and criminals to aid the police. And he was first to forecast the use of infra-red and ultra-violet photography. He also became the first practical documentary photographer when he photographed successive stages of the construction of London landmarks. These included the Royal Exchange, Hungerford Bridge and Nelson's Column in Trafalgar Square.

Before it died out the Daguerre system chalked up some notable firsts. Using the standard silver plates, Doctor John William Draper of New York took the first photograph of the Moon in 1840. His tiny image was gained without a telescope, so it lacked detail. A later daguerreotype (1853) by John Adams Whipple of Boston was the first to show the details of craters on the Moon; but Whipple had a distinct advantage over Draper in the shape of a 15-inch reflecting telescope. This gave him a superb image fully 3 inches wide. In France (1839) Alfred Donne went to the other extreme and produced the first solar-microscope daguerreotype of the eye of a fly. A year later and both lunar and solar eclipses were first recorded. But this magnificent system lacked two essential ingredients. The most obvious was the unique nature of each plate; but added to this was the impossibility of enlarging the image, in part or in whole.

After Talbot all advances were in the refinement of cameras and the creation of new photographic bases. Glass plates replaced the old paper negative material, while the collodion process, first invented in 1851 by Scott Archer, gave a new vigour to photography. Archer's glass plates had to be exposed while the sensitive emulsion was still wet, but despite this handicap, exposures were speeded up and the results were sharp and brilliant. And little was lost in printing the positive. In 1852 Scott Archer devised a brilliant way of using his plates to create 'instant' portraits within minutes. These first *Ambrotypes* were produced by underexposing the glass negative which, after development, was then mounted on a black background. This made the dark areas look light, and the light areas look dark, and a satisfying positive image resulted. The same principle was then used to create the *Tintype*. In its first form, the wet collodion was coated on to black enamelled tinplate. This method, also called the *Ferrotype* process, was first introduced in the USA by J. W. Griswold in 1855. It was then further refined as a dry plate process and special cameras were made, equipped to take these instant pictures exclusively. The cameras incorporated a development and fixing chamber, and delivery of the stabilized picture could by made in under a minute in warm weather. These cameras were mainly for use by wandering photographers and were popular at sea-side beaches, where immediacy, rather than quality, counted. Surprisingly, many of these heavy and clumsy cameras were still in use a century later. Despite the indifferent quality of these direct positive methods, they did at least popularize the idea of having portraits taken.

The general public was thus made ready for the next great step; the introduction of the carte-de-visite portrait. The idea was first patented on 27 November 1854, by the Parisian photographer André Disderi. His method divided the usual large glass negatives into separate sections, the size of a visiting card. Up to ten sections could be achieved on the largest plates, and each of these sections could be exposed, one by one. This was made possible by the use of cameras which had multiple lenses or repeating backs and, sometimes both combined. Each plate when contact printed would produce a number of portraits at one go. This innovation cheapened the cost of portraiture and made it possible for many a poor family to have treasured likenesses on their mantelpieces or sideboards. And the very rich adored the idea as well. In July 1860 the richest lady in Britain, Queen Victoria, commissioned the Regent Street photographer J. E. Mayall, to take cartes-de-visite of herself, her beloved Albert and their children. The resulting portraits were then published as the first authorized Royal collection. This *Royal Album*, first issued in August 1860, led to huge sales of the individual pictures and of complete sets. After Albert's death in 1861, 70,000 copies of his cartes-de-visite were sold in one week. This was the first time that wide-spread mourning had been aided by photography. Photographers soon cashed in on popular curiosity about the great and famous with mass issues of cartes-de-visite of politicians, preachers, writers, inventors, explorers, military men, in short, anyone of fame who had posed for their cameras. For the first time, and long before cigarette cards were thought of, collecting portraits became a widespread hobby. Yet another collecting craze involved the more expensive stereoscopic photographs, first made possible by Sir David Brewster's invention of the stereoscope in 1849. This was first introduced commercially by Louis Jules Duboscq in 1851. Sir David's twin-lens viewer led to the making of the first stereoscopic twin-lens camera and the founding of the first company devoted to stereoscopy. This was the London Stereoscopic Company which first opened up in 1854 and within two years succeeded in selling half-a-million viewers alone. Its photographic sales must have been colossal, since each viewer needed an assortment of prints to make it worthwhile.

Indeed their catalogue came to list 100,000 different cards. This particular craze gained its first great boost when Queen Victoria fell in love with the stereo views shown her at the Great Exhibition of 1851. She became the first monarch to own a viewer and prints, and her example made stereoscopy ultra-fashionable. She little dreamed that the three-dimensional delights of the system made it ideal for exploitation by creators of pornographic photographs. When things leapt out at you the possibilities were endless!

Though wet plates were impressive they had their obvious drawbacks. An attempt to make dry collodion plates resulted in the first process of real merit outlined by Doctor J. M. Taupenot on 8 September 1855. But the first commercial dry collodion plates were based on the rival British patents of Doctor Hill Norris of Birmingham (patented 1 September 1856). All these worthy efforts had the fatal disadvantage of slowing down the exposure speed. This was something not to be tolerated for long, since photographers were hungry for speed. Only faster plates could give them the flexibility they craved. To meet this need at least seven inventors went to work on an improved breed of dry plates. The first to publish significant details of his progress was Doctor Richard Leach Maddox. On 8 September 1871 he gave an account of his gelatine silver bromide emulsion. Historians now hail this as an epoch-making invention, but in its first form it was still far too slow. Others took over the refinement of his emulsion; people like Kennett, Bolton, Sayce, Wratten, Mawdsley, Berkeley, Abney, Charles Bennett, and John Burgess. To these men collectively, we owe the next great practical step in photography.

In April 1878 both Wratten and Wainwright and Peter Mawdsley's Liverpool Dry-plate Company issued their first adverts for the new gelatine coated dry plates. The problem of speed had been solved. For many years to come all plates would depend on the efforts of these pioneers. And their gelatine emulsion is still the basis of those used in present-day photography.

There still remained the problems of weight and fragility. As early as May 1854, A. J. Melhuish and J. B. Spencer had created the first roll-film; but their roll was simply made up from sensitized waxed paper contained on a spool. Ingenious no doubt, but lacking a future. Others did try to make a success of paper rolls, including George Eastman, whose first paper roll-film camera was issued in 1887. This was soon superseded by the first *Kodak* box camera, a simple model to operate with a fixed-focus lens and one shutter speed. It was advanced in its simplicity but still used a paper roll. But it was the first camera to be backed up by a printing and development service, since the camera itself, complete with film, was returned unopened to the Eastman factory. And there the film, with its 100 exposures was dealt with; the camera was reloaded with fresh film and camera and prints were then returned by mail. 'You Press the Button, We Do the Rest', boasted Eastman. Yes, it was all very simple, but very expensive and in 1889 it was replaced by the first camera to take the newly perfected celluloid base roll-film. This first celluloid roll-film (27 August 1889) was monopolized by its creators, Eastman, but as soon as a daylight loading spool was introduced (first in 1895) many manufacturers began to make cameras suited to the new films. This important

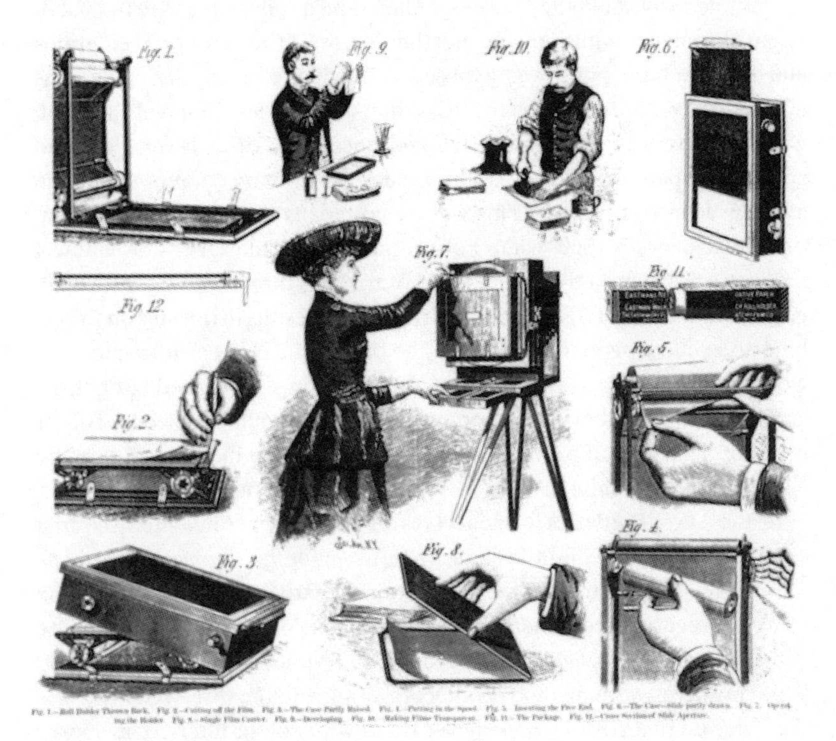

The first roll-films and how to handle them

advance was not an original Eastman venture. The paper-backed film was first created by S. N. Turner of the Boston Camera Manufacturing Company in 1892. But Eastman saw its potential and had the funds and expertise to make it a world-winner. He also saw the value of the Boston firm's *Bull Eye* camera; it used for the first time a red window at the rear to show the number for each exposure. Eastman did not bother to haggle over assets, he simply bought the Boston firm outright and then used the red rear-window as a selling point on the pocket Kodak camera of 1895. His No 2. *Bullet* camera used the window as well and the pair became the first cameras to bring snapshots to the masses. Then the famous box Brownie of 1900 became the camera that first slashed the cost of photography to the bone and was cheap enough and simple enough to be given to a child.

At the turn of the century all plates and films were black and white only. As well they were orthochromatic in scope; in other words they were not sensitive to the entire spectrum. In France the Lumière brothers (Auguste and Louis) set out to remedy this and by 1904 they were able to patent the first one-shot commercial colour plate. Their *Autochrome* plates were sprinkled with a mixture of starch grains dyed red, green and blue-violet. A tacky adhesive held these grains in place and a roller squashed them flat. A panchromatic emulsion was then laid on top of the compressed grains and the resultant plate was exposed through the clear glass side. This allowed the coloured grains to act as filters. After development the plate was re-exposed to light and developed once more. This *reversal* system gave a transparency made up of tiny specks of primary colours, which the eye saw as mixed colours. This was by no means the first colour system, that dates back to 1869, when both Charles Cros and Ducos du Hauron independently worked out a method of making three exposures through green, orange and violet filters and printing the results in register on to red, blue and yellow pigmented papers. The three prints, when superimposed, would then yield up a colour picture. Du Hauron also worked out a three colour additive system using a screen of lines, that could be used for colour printing. This type of printing was first demonstrated in practical form by Charles Joly of Dublin in 1896. His screen was ruled at 250-lines per inch.

In the USA Frederick Eugene Ives of Philadelphia invented the first commercially used colour system when he made his *Photochromoscope* camera in 1891. It took three separation negatives in quick succession using a repeating-back mechanism. The single plate then carried three exposures made through a red, a green and a violet-blue filter. Black and white positives were made by contact printing and the three records were then viewed in a *Kromoskop* viewer fitted with colour filters. The viewer then saw a picture in full colour. In 1895 he even projected stills in colour, but none of his enterprises matched the Lumières' simple single-plate method. And once they had crossed this colour threshold it was not long before others were hard at work on even better methods. The first to better the Lumières' performance was Louis Dufay, a fellow Frenchman. In 1907 he first used a glass plate that was printed with a pattern of narrow, coloured lines. The different colours crossed each other on a precise plan. These lines act as multiple filters and when clamped in register with a panchromatic plate gave an excellent colour rendering. In 1908 the first Dufay *Dioptichrome* plates were put on sale. The method then became known as *Dufaycolor* and was adapted to roll-films and even used for making motion pictures. The quality was fine but exposures were long and the results tended to be on the dark side. These were the major defects of all the subsequent 'additive' systems.

The cameras that produced the pictures were a very mixed bunch. The studio models stayed with large glass plates and black viewing hoods, long after the era of the roll-film began. For that type of camera all major improvements were concentrated on shutters and lenses. Now, a quite simple lens *can* be used for photography, but it will introduce distortions; straight lines will become curved; all colours of the spectrum will not meet at a common focus; the definition will be poor; and it will only work when stopped down and then only slowly. To remedy these defects, designers went for compound lenses at an early stage.

On 1 January 1841 the first corrected compound lens was sold, along with the brass Voigtlander camera made to use it. This fine lens was created by Josef Max Petzval of Vienna and with an aperture of F3, was fully thirty times faster than any other lens of that time. It was great for portraits, since the central definition was sharp, even if the edge definition was rather soft.

This first portrait lens stayed the favourite for the next fifty years (in modified forms it is still used), then it was overtaken by the *Protar* anastigmatic lens designed by Paul Rudolph and first introduced by Carl Zeiss in 1889. An anastigmat is a lens in which all optical defects, including astigmatism, have been corrected by means of compound lenses, carefully calculated curvatures and a mixture of glasses with different refractive indexes. Only an anastigmat can give pin-sharp photographs with accurate details and the 35mm miniature camera could not have come into being without the anastigmat. But the first of the breed of true miniature cameras needed the right type of film to make it work, and that was fortunately provided by the growth of the cinema. Its fine-grained, perforated film was ideal as the starting point for new camera designs.

In 1912 George P. Smith of Missouri made the first still-camera using cine film. Two years later came the first German miniature, Levy-Roth's

(Berlin) *Minnograph*, a camera meant to take fifty pictures on each roll. This was in 1914, the same year that Oskar Barnack completed the prototype of the first Leica camera. But it was not the *Minnograph* that first drew people to small standards, it was oddly enough, a miniature *plate* camera, the *Ermanox*. Made by Ernemann's of Dresden, this first went on sale in 1924. It looked very much like a short-barrelled, giant lens with an oblong box stuck on the back. Its large aperture lenses (F1.8 or F2) made it ideal for use where lighting was poor. Its small size made it ideal for use where conditions were cramped. It was the first camera to alert photo-journalists to the value of miniaturization. So, when the first Leica was put on sale in 1925, there was a small army of converts ready to welcome it. Since it was precision-made it gave ultra-sharp negatives capable of standing up to high degrees of enlargement.

Others, apart from journalists, saw its sterling merits and it became the first successful high-precision camera in the world. It arrived without any real competitors and stayed without rivals until Carl Zeiss introduced his first *Contax* camera in 1932.

After that most firms brought out a miniature of sorts, though not all stuck with the standard 35mm film; Kodak, for one, brought out a new

The Ermanox of 1924.
The first true miniature camera, but it still used plates

unperforated 828 roll-film and built a new camera around it. This *Bantam* camera and the 828 film were first introduced in 1936. In the same year Kodak issued its first 35mm colour film. Known as *Kodachrome* it was only useful for transparencies, and that limited its mass appeal. Also limited in appeal was the first fully automatic camera, first sold in 1938.

This was yet another Kodak venture and a significant one, since this 'Super Kodak Six-20' used a photo-electric cell coupled to the aperture setting and the shutter-speed mechanism. The exposure was automatic and fool-proof. But all this made the camera even more expensive than the Leica, and only 700 or so were ever sold. Now we can see that the camera was way ahead of its time. Yet Kodak made no mistakes when it introduced the first colour print film in 1942. The film was met with popular enthusiasm and became a winner.

Yet not everybody was content with dependency on roll-films, and one discontented three-year-old girl helped bring about the instant-picture revolution. The girl was Jennifer Land and she demanded to see *instantly* a picture taken by her father a mere three seconds earlier. Her father, Doctor Edwin Land of Cambridge, Massachusetts thought 'Why not?' He later said: 'Within an hour the camera, the film and the physical chemistry became so clear that with a great sense of excitement I hurried to a place where a friend was staying to describe to him in detail a dry camera that would give a picture immediately after exposure.'

In November 1948 Doctor Land's brainchild, the first *Polaroid* camera, went on sale in the USA. This Model 95 produced sepia-toned, positive pictures in one minute. Black and white prints soon followed, but the first Polacolor print film had to wait until 1963. Nine years later the old original, peel-apart system was replaced by a single integral picture system. And in 1982 came its first process for producing black and white and colour transparencies instantly.

Polaroid was without competitors until a rival instant picture camera was first launched by Kodak in 1976. It was said to work on a new principle but in October 1985, after a ten-year court battle, Kodak's camera and its films had to be withdrawn and Kodak found themselves faced with a bill for $1000 million damages! But other Kodak ventures were certainly theirs by right, like the Pocket Instamatic with its drop-in cartridge (first sold in 1963); and the Disc Camera with its fifteen tiny negatives and built-in flash. This was first introduced in 1982, and though beautifully simple and efficient never really caught on. Those tiny negatives gave prints that were far too grainy for most people. For all that, it was a brilliant invention, but not even the super-fine 'T' grain emulsions, first introduced by Kodak in 1983, could save the system. It expired in 1988. Few mourners wept.

At present the trend is strongly biased towards cameras that take account of every possible problem. Automatic focus (first used in 1945) is now a priority, as is automatic exposure (first used in 1938), automatic loading, automatic wind-on, and automatic reading of the film speed. The film speed is now read by the DX coding system first perfected by Kodak in 1984. Each roll carries a code which tells the camera its speed and number of exposures.

All films now conform to DX standards, so if your camera is automatic you no longer need a good memory or the end of a packet to keep your light gathering in order. 'I have seized the light; I have arrested its flight!', so said Daguerre when mesmerized by his first image. Now that automation has taken over, some serious photographers are looking back at the old techniques which involved real skill and risks. One Japanese firm is already making mahogany cameras with leather bellows and brass fittings, along the lines of the first field and studio cameras. And there is even talk of reviving the lost skills of daguerreotyping, something that seems like an over-reaction. Yet oddly enough there could be a market for the first daguerreotypes of the twentieth century, and the first man to seriously consider this is, in fact, the author of this book!

• MOTION PICTURES •

The search for pictures that moved and simulated real life began long before the first photograph was developed. It began in the minds of the first men who constructed magic lanterns. Their names are lost but their work was recorded by the Jesuit Anthanius Kircher in his revised *Great Art of Light and Shade* of 1671, by Francesco Eschinardi in 1666, and by Joannes Zahn in 1685. By 1710 the first crude movements were given to lantern slides by Reeves of London, who created such elegant effects as a lively cockroach which could be made to jump into the mouth of a sleeping man. Two slides and a shifting slide-holder animated this bedroom indignity. All later animated lantern slides were little more than refinements on such lines.

We had to wait until the nineteenth century before concentrated thoughts and experiments gave us the first pictures gifted with fluent movement. In 1832 both Simon Ritter von Stampfer and Professor Plateau independently developed revolving disk devices that made pictures 'move'. Think of a gramophone disk with sixteen equal sections ruled out on it. A drawing sits at the top of each section. Each drawing is of the same person or scene but drawn at a slightly different stage of movement. This disk is placed with its face towards a mirror and spun around a central spindle. Sixteen short slots cut out between the pictures allow the viewer to peer through the back. The segments work as shutters, persistence of vision then blends each briefly glimpsed picture with the next to come, and the illusion of life-like movement is created. In von Stampfer's *Stroboskop* and Professor Plateau's *Phenakistiscope* we have two of the essential requirements that are needed for true motion pictures.

In 1845 Lieutenant Franz von Uchatius, of the Austrian army used von Stampfer's ideas but placed the pictures on a glass disk and revolved the disk in a magic lantern. He was able to project moving figures on a wall but they were small and dim. His lighting and narrow shutter-slits were unable to give large, sharp images, but they were the first projected moving pictures. Then, in 1853 he added a powerful limelight; eliminated the shutters and caused the light to revolve and illuminate (one by one) twelve pictures placed behind twelve lenses; each lens capable of being angled at a screen so that its image coincided with all the others. This was the first attempt to create a moving picture teaching-aid, for that was the task set him by the Austrian General Staff.

Uchatius, like all his fellow experimenters, was completely dependent on hand-drawn images, but with the birth of photography a new realism beckoned. In 1860 the Frenchman Pierre Desvignes took 'action photographs' of a stationary steam-engine and mounted them on a Plateau disk viewer. These first moving photographs were not quite what they seemed. Each photo had been taken with the steam machine at rest, and the machine parts had been moved by hand into each new position. But Desvignes was not cheating, it was simply that he had no other way of working. His camera was not adapted for rapid work, his photo plates were too slow, so he had to resort to stop-motion.

In the USA, Henry Renno Heyl used this stop-motion technique to create some impressive sequences featuring a couple waltzing and an acrobatic act. On 5 February 1870 he screened his 'moving pictures' at the Academy of Music, Philadelphia. The audience is said to have 'comprised 1,500 persons'. But did they all watch at the same time? He used a variant of Uchatius' first machine to project his eighteen picture disks at this first ever projection of moving photographs.

The first man to photograph an unposed sequence of pictures of movement was Eadweard Muybridge, an Englishman living in the USA. In 1872 he is said to have been commissioned by a former governor of

Eadweard Muybridge, pioneer of stop-motion photography

California, Leland Stanford, to photograph a horse in motion so that a bet could be settled. Were all four hoofs off the ground at any time? This momentous question could be readily answered by photos taken by a row of cameras placed alongside a straight course. Muybridge used twelve matched cameras placed 21-inches apart and housed in a long, solidly built shed. Twelve threads, one to each camera's shutter, were stretched across the track and were broken, one by one, as the horse ran past. Each broken thread operated a shutter and twelve fine pictures resulted. Later on, the camera line-up was doubled for even more precision and people were photographed in action as well. There has been some dispute over the date of Muybridge's first public projection of his photographs of real-life motion.

But one date is certain, since the *Scientific American* reports a 4 May 1880 meeting of the San Francisco Art Association at which Muybridge used his *Zoogyroscope* to throw upon a screen ' . . . apparently the living, moving animals. Nothing was wanting but the clatter of hoofs upon the turf.' In 1893 Muybridge displayed his work – by then extensive – in the first hall specially built for viewing moving photographs. This was the Zoopraxographical Hall at the Chicago World's Fair.

But Muybridge acted only as an inspiration to the pioneers of the cinema. His method was admirable but never the real answer. Those who guessed at the answer were looking to pictures mounted on a band or printed on a strip. And they knew that only many pictures each second could give a natural flow to action. The first outstanding steps towards movies were taken by Frenchman Louis Aimé Augustin Le Prince. Aged twenty-five he married the daughter of manufacturing brassfounder, John Whitley of Leeds, and came to England. There, in Leeds he began to study photography. His training in chemistry and his work as an artist well suited him to master all things photographic. Work took him to the USA in 1881 and while there he became involved with Poilpot's Panorama, where huge canvases were painted and erected to give a breathtaking impression of a battlefield, a mountain range, or a soaring cathedral. But despite the attempts at realism everything depicted was doomed to stay static. Even unrolling a scene inch by inch was unconvincing. Le Prince began to think of moving pictures in full colour projected on to the canvas. This would outstrip any fixed painting in glory and satisfaction, especially if the images could be made three-dimensional. He began thinking of a camera to make all this possible.

On 2 November 1886 Le Prince filed his first patent application in Washington. Like most inventors he had a number of different devices in mind, but out of the many possibilities emerged a single lens camera, with a means of taking pictures at a rapid pace, but with each frame *halted* for the taking period; in short, with an intermittent movement. But did it work? The evidence of those film strips that survive says, yes. Le Prince's biographer Christopher Rawlence has displayed the footage in his documentary film *The Missing Reel*. Each strip of twenty frames gives naturally paced action and shows that the camera of 1888 was the first ever to take realistic motion pictures. This must make Le Prince the undisputed father of cinematography. But his labours were never taken to full victory. He mysteriously vanished on a trip to France in 1890. He boarded the Paris train at Dijon and was never seen, dead or alive, again.

By contrast the frames left behind by William Friese-Greene show that the action was recorded at too slow a speed to give natural results, possibly as slow as three or four frames a second. The romantic view of him, embodied in the film *The Magic Box*, as a neglected, unhonoured genius and father of the movies, is at variance with the truth. His claim to be first is wishful thinking, and no more than that. But he was an endearing dreamer of dreams, always coming close to solutions, but never quite making it. For this, he deserves to be remembered.

Le Prince was greatly hampered in his experiments by the lack of a strong, transparent band able to carry his pictures. Sensitized paper bands were the best the photographic industry could offer. This all changed in 1889 when the first celluloid-based roll-films were issued. Celluloid had been around for some years before its immense value for photography was fully realized. In fact, the first patent for celluloid was granted on 15 June 1869 to John W. Hyatt, of Newark, New Jersey, though it was then called *Pyroxlin*. It was a combination of collodion with camphor, originally intended to be used as a substitute for ivory. It was first called by the name *Celluloid* in the *US Patent Gazette* on 2 July 1872. In May 1887 Hannibal Goodwin applied for a US patent on a method of preparing a celluloid support for photographic emulsions. This is the first mention of such a vital combination, though Goodwin himself was never engaged in the manufacture of his patented films. The first man to make such films was John Carbutt of Philadelphia. His first efforts were made in 1886 and his first commercially available sheets were offered in 1888. On 9 April 1889 Harry Reichenbach applied for a US patent on his method of making transparent film bases out of celluloid. It was this patent that the Eastman Dry Plate Company took over and it was the first batch of this product that was sent to the Edison Laboratories late in September 1889.

It arrived at just the right time. Edison's employee, William Kennedy Dickson, was then brooding over the idea of adding pictures to a phonograph. In other words he was dreaming of making *talking-pictures*. Most of the accounts of what followed are contradictory and hardly believable.

Dickson is supposed to have created talking-pictures by the fall of 1889 and projected them on to a 4-foot screen. This is claimed as a double-first. The truth is that in 1889 he was still trying to work with tiny pictures arranged around a cylinder and linked to a phonograph. This device was hardly more than a peep-hole machine and could never serve as a projector. Furthermore, this cylinder machine was soon dropped in favour of a very different machine, the *Kinetoscope*, so it is doubtful if it was ever of much use. The new Kinetoscope, by contrast, did work and it provided the first ever publicly viewed realistic motion pictures. Even so, these were not screened pictures, they had to be viewed as a peep-show, through a magnifying glass. The viewer was entertained by action recorded on an endless 40-foot band of film, and a revolving slotted disk acted as a shutter and gave the images fluidity. This Kinetoscope was first openly demonstrated at the Brooklyn Institute in May 1894. But the first large-scale shipment of these machines did not reach their showplace at 1155 Broadway until 14 April 1894.

The viewing public was entranced. For a quarter they could see the films housed in five machines. Subjects included wrestlers, trained bears and staged comic episodes in barber-shops and laundries. And in these

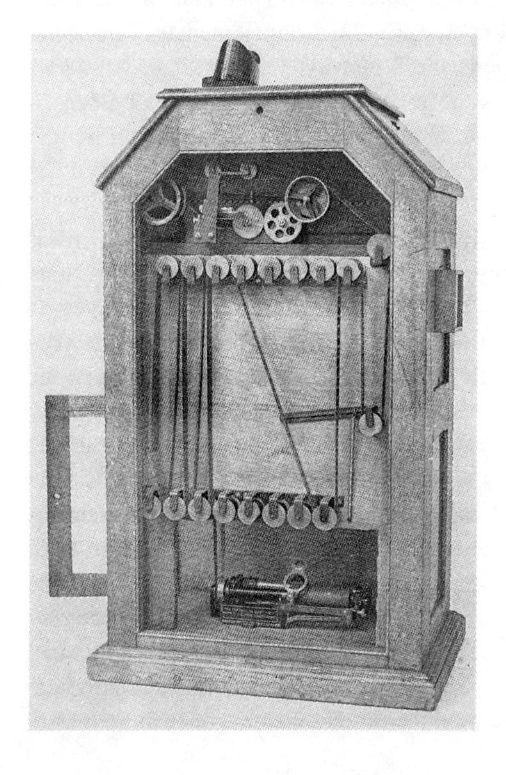

A Kinetoscope, the first commercial motion-picture machine

first film arcades they saw their first 'Stars', for Dickson had engaged Sandow the Strong Man, Buffalo Bill, and Annie Oakley to pose for his camera. All these entertainments were filmed in the first film studio in the world. It was specially built, with a roof that opened up to let in sunlight. And the whole studio was mounted on a circular track, in order to keep pace with the sun.

Dickson organized everything in this studio and became the world's first film producer. But his talents never developed any type of film projector. Some reference works show Jean Acme Le Roy as the first man to develop and use a film projector. And an advert dated 22 February 1895 has been reproduced as evidence of Le Roy's primacy. This advert, relating to the Opera House, in Clinton, New Jersey certainly looks convincing. But research by film historian Gordon Hendricks, has shown that no opera house existed in Clinton at that time. The advert may well have originated as a deliberate fake to be used in the court battles waged by those producers excluded from the early 'movie trust'. There is no dispute, though, over the date and place of the first film to be shown publicly on a screen. It was screened at 44 rue de Rennes, in Paris on 22 March 1895. The film showed workers leaving the Lyons photographic factory owned by the Lumière brothers (Auguste and Louis), and this first Lumière production was shot in September 1894. Their first audience was drawn from members of the Société d'Encouragement à l'Industrie Nationale. In the same year Major Woodville Latham demonstrated a rival projector in an empty store on Broadway. This screening (on 20 May 1895) was of a four-minute film of a boxing match. It was a limited menu but it did attract a paying audience, so it counts as the first showing ever to do so.

An earlier, first demonstration of Latham's machine on 21 April was free, but only open to the press. This event angered Edison, who came up with the wild boast: 'In two or three months we will have the Kinetoscope perfected, and then we will show you screen pictures. The figures will be life size, and the sound of the voice can be heard as the movements of the figures are seen.'

This Latham projector was of importance for one famous feature, it used extra sprocket wheels to handle a loop of slack film. This loop allowed the film to stop and start in the projector gate without tearing. Earlier machines needed to drag the whole reel of film at each stage of its cycle. This meant tiny reels and short satisfaction. The loop made large reels possible. Yet another projector of 1895, T. Armat's *Vitascope*, made use of a loop of film, but used a different mechanism to activate it. At an early stage a machine was taken by Armat to Atlanta, Georgia in September 1895 and there he obtained a concession from the body controlling the Cotton States Exposition. The concession allowed him to build a movie theatre in the Exposition grounds. It was the first theatre of its kind, but there were no queues controlled by doormen, for the Exposition failed to draw crowds and Armat ended up with heavy losses. But he did become the first full-time professional projectionist when he opened a season of films at the Koster and Bials Music Hall, in New York City, on 23 April 1896. He used his perfected projector for the first time, though (for publicity reasons) it was described in the press as an Edison machine. Press reports also reveal that one of the films shown was a hand-coloured short called *Umbrella Dance* this counts as the first film in colour ever shown. The same report states that the short entitled *The Band Drill* was of a scene taken from Charles Hoyt's Broadway success *A Milk White Flag*. This was the first time that a stage play provided material for a film.

In Britain the first motion pictures were screened by Birt Acres at the Royal Photographic Society headquarters, 14 Hanover Square, London on 14 January 1896. He showed films taken by him including *The Derby*, *Rough Seas at Dover* and *The Opening of the Kiel Canal*.

In the next month, on 20 February the public had its first chance to pay to see movies. The show was held at the Regent Street Polytechnic for a three-week period. The tickets cost one shilling and the audiences were treated to a selection of films made by the Lumières. Presentation was in the capable hands of a French magician named Félicien Trewey.

1896 was certainly the first great year for the cinema. There were first screenings in the great cities of Turin, Vienna, Oslo, Dublin, St Petersburg, Johannesburg, Budapest, Madrid, Bucharest, Belgrade, Copenhagen, Lisbon, Helsinki, Bombay, Ottawa, Buenos Aires, Rio de Janeiro, Shanghai, Mexico City, Melbourne and Auckland; all in that order. And lesser venues in the Netherlands, Sweden, Czechoslovakia and Poland had screenings as well.

Those early viewers were grateful for anything that moved and this resulted in some alarming trends when film makers began to intercut real news footage with studio footage. Even worse was the faking of complete films that posed as authentic news. This deplorable practice was first indulged in by the Lumières when they created their film depicting the trial, conviction and exile of Dreyfus. Not one frame was authentic.

But authentic news was recorded faithfully by other companies. As early as September 1896 the campaign of William McKinley had been covered; this was the first filming of a US Presidential campaign. The 4 March 1897 inauguration ceremony was covered as well, giving us the first filmed record of this important ceremony and the first shots of a US President (Grover Cleveland).

Sadly, the news footage of the Spanish-American war seems mainly to be bogus, which means that the first honest films of men at war must be those shot during the Boer War of 1899–1902. These films include the surrender of Boer General Cronje to Lord Roberts; the first footage of the end of a major war. The earliest reel of this war was filmed on 12 November 1899 by John Bennett Stanford and is still preserved in Britain's National Film Archive. (Earlier footage, first shot during the Greek-Turkish conflict of 1897 by Frederick Villiers, was never shown or preserved.)

On the openly fictional side there were films that took pride in fakery, like George Albert Smith's *Corsican Brothers* of 1898. The brothers in the original story were twins, but twin, sword-wielding actors were hard to come by, so Smith invented the first movie double-exposure method to allow one actor to play two parts simultaneously (or so it seemed!). Smith has other early firsts to his credit. He used the first close-up in a British film in a 50-foot reel called *Man Drinking*, in September 1897. He intercut close-ups with other scenes in his *Grandma's Reading Glass* of 1900. And he used the first known 'wipe' in his *Mary Jane's Mishap* of 1903. His biggest triumph, with colour came later.

The first dissolves from one scene to another were accomplished by Frenchman George Méliès in his *Cendrillon* of 1889, while the first fade appears in James Williamson's English film of 1904 *The Old Chorister*. So most of the important techniques were hit on within the first eight years or so.

But technical skills and novelties were not enough to give real momentum to the industry. The first documentary of realistic length and treatment was made as early as 1897. This French footage ran for twenty minutes and recorded the activities of *The Cavalry School At Saumur*. But this was exceptional and most items were brief to the point of exasperation. Such shorts could only satisfy for a limited time. The newborn 'Kingdom of Shadows' needed rattling good yarns and real-life adventures to hold its subjects enthralled. The first adventures of quality and meaning were first brought to the screen by the American Charles Urban. In 1903 his cameramen were sent off to a range of visually exciting places, including the Alps and Borneo. They returned with rich material for the first 'travelogues'. At the same time Urban commissioned the first films showing the mysterious terrain that only the microscope could explore. These films, taken through a microscope by F. Martin Duncan, were released under the title *The Unseen World*. They gave the masses their first chance to marvel at a world that for too long had been known only to a tiny minority.

But the fictional narrative film was slow to develop into a satisfying form. French showman Georges Méliès had tried his best but his vision was limited. He relied too much on the theatrical conventions he knew so well. His camera was a recording eye placed in front of a stage. Even when he departed from this static mode he did little more than move people and objects *towards* his fixed camera, treating it like a solitary spectator. Despite this he was the first film-maker to draft a scenario, and the first to increase the length of fiction films. His rivals were offering mere snippets compared with the 1,414-foot length he reached in his *Impossible Journey*.

The Méliès legacy crippled the first attempt to raise the standards of story-telling. The French *Film d'Art* movement used fine, seasoned actors and the first scenario written by a noted playwright, but it offered little more than a film of a stage play. Its *Assassination of the Duc de Guise*, written by playwright Henri Lavedan was acclaimed by high-brows. And it featured the first film music written by a major composer, Saint-Saëns. It was a great success, but led to a cul-de-sac. Film d'Art productions were stagey, yet without the popular charm of the Méliès reels. The movement failed to grasp the fact that film had its own language, its own techniques and its own, new-style audiences. These audiences were looking for longer, finer stories with excitement and

George Méliès, master of fantasy

grand climaxes. Edwin S. Porter tried to provide for these needs and his *Great Train Robbery* of 1903 became a box office success because it offered 'Thrilling and exciting incidents in fourteen scenes'. It was the first important 'Western' and the first step upwards for the American film industry. It was based on a continuity of action that gave a sense of mounting excitement. It exploited the ability of the camera to make events dynamic as its actors moved towards and away from the lens. In its last scene a bandit even thrust his pistol at the camera and pulled the trigger. The audience ducked.

All the first 'Westerns' were shot in the eastern parts where the early companies were located. The move to the West Coast was a tardy one, for the first film studio in Los Angeles, that of William Selig's company, was not opened until 1909 at Edendale. The first studio in Hollywood itself, the Centaur Company's, only started up in October 1911. But it was an historic opening since Centaur's Hollywood branch (Nestor) became the first company to initiate the great film industry invasion of Hollywood. Due to the exceptionally fine weather it enjoyed, Nestor Studios was able to turn out a steady stream of pictures, all of high photographic quality. Those out East were impressed and hurried West to look at Arcadia. Within months, fifteen more film companies were working in the Hollywood area and the industry's future centre of gravity was determined.

It was D.W. Griffith who made the first film in Hollywood. His *In Old California* was begun on 2 February 1910 and finished two days later. For Griffith this was just one of the many productions that allowed him to experiment and bring together the many techniques and insights that he then used in his masterpiece *The Birth of a Nation*. His *Birth* was the culmination of five years of intensive, hard work during which he fathered over 455 movies.

The Birth of a Nation has been described as the 'first picture *conceived* as more than a silent film'. The editing had been harmonized with a great many musical themes, from different sources. The pace had been musically guided at each scene. But this was not the most obvious factor that makes this the first great feature film. It was outstanding because Griffith amalgamated all the filmistic wisdom that had evolved by that time. He refined the art of editing, making an inspired use of parallel cutting and dynamic switches from action to action. In many ways his treatment was melodramatic, but it suited his material and the psychology of the audiences of that time. Yet there were subtle, poignant moments as well, often captured in bold close-ups. Added to all that was his painstaking pre-planning of every scene and his comprehensive rehearsals. All this made his *Birth* the first film to break completely with the theatre. It is regarded as the first to prove that film was an independent art. Today, its Civil War, Confederate sentiments, with its approval of the Ku Klux Klan, makes it an unpopular film to screen. And existing copies are faded and often shown at too brisk a speed. Remember, as well, that it was originally shown with a specially written, enhancing score, performed by a full symphony orchestra. This makes an objective re-appraisal difficult for some, but the magic is all there, just waiting its chance – even without music.

Alongside the *Photoplay* (a name of dignity!) the industry began to develop the cartoon film. The first cartoon can be accurately dated by the copyright frames inserted by the Edison Co. It was J. Stuart Blackton's *The Enchanted Drawing* of 16 November 1900. It was a mixture of live action and drawings in which an artist (Blackton) draws a face that comes to life and responds to each change made by the pencil. A similar theme was used in the first British cartoon of 1906. In *The Hand of the Artist*, Walter Booth cast himself as the artist whose pencil brings to life a coster and his girl. They cake-walk and enjoy their brief period of liberation. This artist and his creations theme was exploited by most of the cartoon makers, and no one attempted to tell a story until 1908. The first of the story cartoons was Emile Cohl's *Fantasmagorie*, first shown at the Théâtre du Gymnase in Paris on 17 August 1908. Cohl then created the first cartoon series, based on his character Fantoche, a matchstick man at odds with the world. His debut was in *Le Cauchemar du Fantoche* of 1909. In the USA the first cartoon series was released by Pathé in 1913. It was a product of the Bray Studios of New York entitled *Colonel Heeza Liar in Africa*. Colonel Liar then became the first American regular cartoon character.

John R. Bray, co-owner of the Bray Studios, was the first patentee of the method of using translucent sheets carrying background drawings. He also patented a method of registration of each drawing, making camera work that much more accurate. And his patent made the first mention of drawings on celluloid, a labour-saving device, since it meant that only the moving parts had to be redrawn on each successive frame. This master patent was first applied for on 9 January 1914. Even finer ideas were then set out in Earl Hurd's first patent of 19 December 1914. He proposed opaque backgrounds over which celluloid sheets were registered. These sheets carried the action drawings. This was the winning method of making cartoons, the one still used today. And its importance was recognized by Bray, who joined up with Hurd and formed the first modern-style cartoon outfit, the Bray-Hurd Company, in 1917.

All cartoons were pure fantasy until Jewel Productions (USA) made the first dramatic cartoon, *The Sinking of the Lusitania*. It was released on 15 August 1918 and according to *Motion Picture News* was the result of twenty-two months of work involving 25,000 drawings made by famous artist Winsor McCay. It was the first cartoon of such length and the first made as war propaganda.

From 1916 on there emerged a growing army of new cartoon characters, but the greatest of them all came late on the scene. Mickey Mouse was first seen on 18 November 1928 when his *Steamboat Willie* was premiered at the Colony Theatre in New York. Oddly enough this was not the first Mickey film. Before it came the silent production *Plane Crazy*, but this was put to one side in favour of *Steamboat Willie* for *that* had sound. In fact it was the first Disney cartoon to have sound. Disney was a great enthusiast for sound in the cinema and the first of his musical offerings, his first *Silly Symphony*, was first shown in July 1929 at the Carthay Circle in Los Angeles. Three years later, on 15 July 1932, he showed his first *Silly Symphony* in colour *Flowers and Trees*. This was screened at the famous Grauman's Chinese Theatre in Hollywood and was the first ever cartoon to employ the Technicolor three-colour imbibition process. An earlier colour cartoon, the sequence introducing *The King of Jazz*, was a Technicolor production but used the earlier two-colour process. It was shown on 30 March 1930 which makes it the first imbibition-printed cartoon to be seen. But it was simply an introductory item, not meant to stand on its own; so the first self-supporting colour cartoon has to be Ted Eshbaugh's Multicolor production *Goofy Goat*. This was first released at the Loew's State Theatre, Los Angeles, on 2 March 1932 but was first shown at a preview at Warner's Alhambra Theatre, on 6 July 1931. *Goofy Goat* is notable for yet another first; it was a 'talkie', the first in colour. Not the first cartoon 'talkie' though. That record belongs to the black and white film *My Old Kentucky*. This Max Fleischer production of 1925 had a sound-on-film track created by De Forest's Phonofilm Labs. A very forward leap.

Yet not every forward leap resulted in triumph. It is doubtful if the usually listed first colour cartoon, *The Debut of Thomas Kat* (Bray Productions, 1916) was a success. It used the Brewster natural-colour system, but in 1916 this system was dogged by technical problems and gave uneven results. This is probably why no other Brewster Color cartoons were attempted.

But colour and sound were essential for satisfactory realism, and as one colour system faltered, so another arose to overtake it. It was very much the same with sound. Edison's former assistant, William Kennedy Dickson, has laid claim to be the first to add sound to screened motion pictures, and he puts the date as 6 October 1889. According to him, on that day, he showed a talking picture to Edison. The figure on the screen (Dickson himself) raised his hat, smiled and said, 'Good morning Mr Edison, glad to see you back. Hope you like the kinetophone. To show you the synchronization I will lift up my hand and count up to ten.' This is a grand story, constantly repeated, yet it is nothing but direful fiction. In 1889 Dickson was still struggling with a worthless device based on a large cylinder covered with a spiral of tiny photographs. These images could only be viewed through a microscope eyepiece. Projection was out of the question.

It was not until 1899 that the first US patent was taken out involving a synchronized system combining the phonograph and the cinematograph. This was the work of George W. Brown. But prior to this Oskar Messter had actually presented the first synchronized sound films in Berlin in September 1896. Unlike the Brown and Edison ideas, his system depended on the flat disks of the Berliner Gramophone Company. The disks were most probably those of singers. Talking pictures, as such, were first shown by the Gaumont Company in its Phono-Cinéma-Théâtre sited at the Paris Exposition of 1900. On 8 June Gaumont showed footage of extracts from five plays, including *Hamlet.* The artistes included Sarah Bernhardt and Coquelin. In the same year, in Britain, Bio-Tableaux Films were displaying films of music-hall performers singing their hit numbers. This first 'hit parade' included songs from Vesta Tilley and the maudlin *Blind Boy*, sung by G. H. Chirgwin.

Within a few years Britain saw the introduction of a number of ingenious synchronizing systems. In the *Cinéphone* method first used in 1908, the overworked projectionist had to look to his reels; spot a photographed clock face on screen; watch an illuminated pointer at the base of the screen and wind a handle to keep the pointer and clock hand in exactly the same position. One word sums up this system: frantic. An easier system was Cecil Hepworth's *Vivaphone* (No. 28, April 1910) this only obliged the projectionist to keep a pointer upright to hold gramophone and film in synch. This system was used to record the first synchronized movies of leading politicians making speeches. The first to record was Bonar Law, followed by Lord Birkenhead. Several other Cabinet Ministers came along to court publicity and it was finally agreed that Hepworth could make the first films ever of a full Cabinet meeting at 10 Downing Street. When the great day dawned the Cabinet Room was rigged out with arc-lamps and Hepworth and his crew stood by for the historic session. Then came the message that deflated everyone – the filming had been called off at the last moment. As Hepworth explained, 'Somebody got hold of the knowledge that members of the Cabinet were to be filmed. Somebody else ... saw only the funny side of it, and how easily it could be ridiculed. That sense of humour ran riot through the newspapers and the British public laughed. Cabinet Ministers do not like laughter ... a great opportunity lost, killed by ridicule.'

Were they really that pompous? Still the loss was not all that great for all these early systems lacked conviction, even when everything kept in step, for the sound was far too feeble. To remedy this Sir Charles A. Parsons invented the first effective sound amplifier, the *Auxetophone*, in 1906. It used a principle first patented by Edison in 1878 (No. 1644) when he devised his *Aerophone*. This was a method of using compressed air to magnify sound from a recording or through a megaphone. Parsons perfected these early ideas and used an electric motor to feed compressed air to a valve controlled by the movements of the gramophone playback needle. This produced a great increase in volume, but the constant loud hiss that accompanied this made the device unwelcome in the average home. But it was ideal for a large cinema and until electronic amplifiers came along it was the only means of giving sound equality with vision.

For others the real solution to the twin problems of amplification and synchronization lay in the development of new *electrical* methods of recording tracks and boosting sound. The great pioneer of our present-day sound-tracks was a French inventor who lived and worked in South London. Eugène Lauste was once on Edison's staff. When he left Edison in 1892 he designed a petrol engine but discarded it when he was told that such engines had no future. So the motor car industry lost his genius and the film industry gained. He went on to make projectors and cameras for Major Latham and, later, for the American Biograph sound-on-film movie house. In the same year De Forest made his first British sound-on-film talking picture. This was a comedy short, *The Gentleman* made at his Clapham Studios, the first studios to employ commercial sound-on-film recording apparatus. Ironically his studios were just minutes away from the original place where neglected Lauste made the very first sound-on-film footage.

Not everyone thought that sound-on-film was the answer. Warner Brothers Pictures opted for *Vitaphone* sychronized disks as the way forward. Sam Warner had been converted by a demonstration at the Bell Labs. He reported, ' ... when I heard a twelve-piece orchestra on the screen ... I could not believe my own ears. I walked in back of the screen to see if they did not have an orchestra there synchronizing with the picture. They all laughed at me.' This visit resulted in the first historic contract between Warners and Western Electric (owners of Bell Labs) signed in June 1925. The first Vitaphone pictures were planned as high-class concert items using the New York Philharmonic Orchestra and

The first amplifier was Parson's 'Auxetophone'

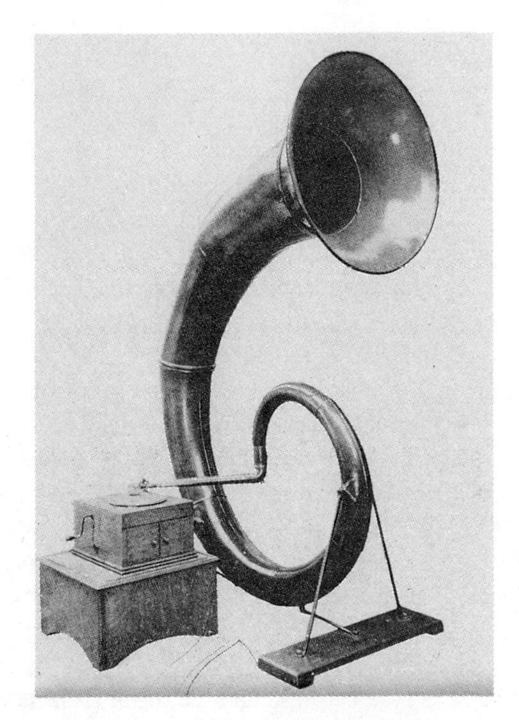

famous soloists. At the same time it was decided to add a musical score to their current silent production *Don Juan* – a romantic drama starring John Barrymore. But their first special sound studios, Vitagraph Studios in Brooklyn, had to be spurned within months, since the rumble of subway trains was picked up by the microphones. A far finer recording venue was found at the Manhattan Opera House and their first usable Vitaphone movie disks were cut there.

The great Warner launch took place at their Warner Theatre on Broadway on 6 August 1926. The very first speech heard by the audience was made by Will H. Hays, former Postmaster-General and then president of the Motion Picture Producers Association. *The New York Times* said of this, 'It was the voice of Hays, and had any of his friends closed their eyes they would have immediately recognized his voice.' Within two weeks this show had set an all-time record when it grossed $29,000 in a single week. The first time any show had attracted so much revenue in seven days.

The second Vitaphone programme, of 5 October 1926, featured Al Jolson for the first time. He sang three songs including *April Showers* and *Rockabye Baby with a Dixie Melody*. The audiences adored his presentations but, more importantly, the Warners shrewdly saw him as the only man to star in their projected *Jazz Singer*. The lead was to have been given to George Jessel, but it went to Jolson. *The Jazz Singer* opened at

Warner's on Broadway on 6 October 1927. Its now legendary impact came about because Jolson ad libbed before two of his numbers. The remembered lines were 'Wait a minute. Wait a minute. You ain't heard nothin' yet! Wait a minute, I tell you. You ain't heard nothin' ... ' Nothing poetic or noble about these words, they were simply the lines he had used on stage after most numbers, but they struck a chord with the audience. And the other, longer ad-libbed sequence in the film convinced them that easy, natural talk was exciting and desirable. And Vitaphone was there, eager to cash in on the new passion.

Other movie moguls were slow to follow up the trend, but keen enough to find ways around the Vitaphone monopoly. William Fox of Fox Films had earlier invested in De Forest's research, but the two men quarrelled and Fox then began to cultivate a similar sound-on-film system developed by Theodore Case and Earl Sponable. But first Fox rented Vitaphone equipment in order to lay hands on the Western Electric amplifying equipment. He used the Case-Sponable system for his first Movietone programme shown in New York in January 1927. That first footage was of a series of songs sung by Raquel Meller, but behind the scenes Fox had plans for the first talking newsreels. And on 2 May 1927 he presented the second selection of Movietone shorts, and there were the newsreels. The audience at New York's giant Roxy found the real-life action sensational. Army bands blared out, West Point cadets thundered past as they drilled. And to cap it all there was Lindbergh taking off on his record-setting flight across the Atlantic. *Variety* reported, 'The roar of the motor when it starts, and the following mechanical stutterings and stammerings as the engine begins to drag its load thrilled everybody in the house.' Fox followed this by screening the first sound footage of Lindbergh's hero's welcome home at Washington DC. This showing on 14 June 1927 also included a piece in which Mussolini made a brief speech, the first-ever film recorded by a dictator.

The battle lines of the two very different systems were then drawn. The problem of colour, though, was still unresolved even after thirty or more years of thought. As early as 1896 Robert Paul had actually hand-coloured every frame of one of his films, some 12,000 tiny pictures in all. By 1910 many machine-stencilled colour shorts were being churned out by the method first developed by the French Pathé Company. In other cases whole scenes were toned by chemicals, or dyed to create an overall mood – red for fire sequences, green for a pastoral effect, blue for a mountain episode. But none of these systems depended on the camera.

The first natural colour process of merit was devised by Edward Turner with the financial backing of F. Marshall Lee. The Lee-Turner patent (No. 6202 of 1899) describes a camera able to record red, green and blue on successive frames. Each exposure was made through a segment of the camera shutter fitted with the appropriate colour filter. The reasoning behind the patent was flawless, but the projection system was not efficient enough to make the venture a commercial triumph. The triple-lens projector was supposed to superimpose the three images and make them one, but the images jiggled around and refused to register properly.

The first commercially successful natural colour process was a by-product of Turner's system. The three-lens projector was replaced by a single lens model and the three-segment camera shutter was replaced by a two-segment version carrying green and red filters. In action, one frame would record all the reds and yellows, while the next recorded the yellows and greens and a little blue. Taking speed was thirty-two frames per second, twice the standard speed. The projector was fitted with a revolving shutter carrying red and green coloured glasses, so that red and green frames were projected one after the other through the right filter. The eyes did the rest, by blending the two images together to give an impression of a wide range of natural colours. But to make all this possible, George Albert Smith had first to develop a sensitive film-coating that was able to record green and the red end of the spectrum. After many failures he was able eventually to work out a way to bathe his negative film in a specially compounded sensitizing dye solution. This gave the results he was after and makes Smith the first man to develop a panchromatic film stock. But it was never marketed. All supplies were kept for the exclusive use of the new system, which was named *Kinemacolor* and first demonstrated in July 1906.

Its first Press-Showing was on 1 May 1908, but the first public presentation to a paying audience did not take place until 26 February 1909. The first cinema to show the new wonder was the Palace Theatre, Shaftesbury Avenue, London. This first public performance consisted of twenty-one short documentary films – more like fragments of a newsreel than anything else – while the first dramatized film in this first natural colour system was *Checkmated,* released in 1910. From the beginning this innovation intrigued the Royal Family. King Edward VII and Queen Alexandra first visited the private showing at Knowsley, on 6 July 1909. A Command Performance was staged at Sandringham on 29 July 1911 before Queen Alexandra. And on 14 September 1911 the first screenings were made at Balmoral.

For the general public, though, there was little choice of time and place. All these wonders could at first only be seen at one place in London, then on 24 March 1910 Kinemacolor was first shown in the provinces, at Nottingham and Blackpool. The death of King Edward VII provided a breakthrough since the funeral was covered by the new system, and everyone wanted to see the event in colour. In May 1910 this first royal funeral in colour was seen in Glasgow, Derby, Nottingham, Blackpool and Burton-on-Trent.

The first Kinemacolor cinema was established at the Scala, off Tottenham Court Road in London. It staged its first performance on 11 April 1911 and in the next few months showed the first Coronation procession in colour (June); the Naval Review; and the Investiture of the Prince of Wales at Caernarvon (July). Then the company made film history when it took its cameras out to India and shot the first colour footage of the country itself and of the many pageants involved in the Delhi Durbar of 1911. This Indian footage was first shown at the Scala on 2 February 1912. It ran for two and a half hours and was the first film to reach such a length. It also impressed the *Morning Post* which wrote, 'It is quite safe to say nothing so stirring, so varied so beautiful, so stupendous, as these moving pictures, all in their natural colours, has ever been seen before.'

Alas, such praise was never given to its subsequent productions. They were mainly photoplays, like *Mephisto*, *Faust*, *Robin Hood* and *Dr Jekyll and Mr Hyde*, but despite the choice of dramatic subjects, the 'acting was poor and the direction was worse'. Without films of impact to show, the company went into decline. Patent battles in the courts helped it on its way down and by 1917 the system was abandoned. But in its time Kinemacolor had made viewers eager for true colour and this spurred on the inventive to greater efforts. It hastened the advent of finer methods. And this interest was raised on an international basis, for Kinemacolor had been shown in Paris (July 1908) in the USA (11 December 1909), in Italy (1912) and in Japan (September 1913). The Japanese venture was especially notable since it opened seven Kinemacolor theatres in Tokyo alone and ran them up until the First World War, when production of new films almost came to a halt.

The first important attempt to replace Kinemacolor was made by a company set up by three Boston engineers: Doctor Herbert Kalmus, Daniel Frost Comstock and W. B. Westcott. Using patents by Westcott and Comstock, they formed, in 1915 The Technicolor Motion Picture Corporation. At one stroke they launched the colour system that would, in the end, revolutionize the industry and they gave the dictionary its first invented adjective derived from film-making.

Technicolor started as a two-colour additive method, projecting the film through colour filters, as in Kinemacolor. Their first film made in this fashion was *The Gulf Between* of 1917. It taught the company so many

Trade advert for the first 'natural choice' system

lessons that this first method was smartly dropped. An alternative method then took its place. Now, the two records captured by the camera (through a beam-splitting prism) were printed on to two films. Each film was dyed, one red-orange, the other green-blue, and the two films were then cemented together to film one record. Amazingly this worked and gave pleasing results. The first film in this new additive Technicolor was *The Gulf Between* first shown at the Aeolian Hall New York on 21 September 1917. It was also the first full-length colour feature produced in the US. But although this new approach worked it developed problems. The double thickness of film and the stiffness added by the cementing medium, made the film cup as it went through the projector's gate. All minds at Technicolor then looked for a new solution, a third system. They developed a camera able to take three images at the same time. Two prisms cemented face to face intercepted the beam of light from the lens and sent one third of the light sideways, to a film sensitive to blue and allowed two thirds straight through to two films sandwiched together. Top film in this bi-pack recorded the green and passed red light through to expose the second film. All three films ran together in perfect synch and provided master negatives for an amazing printing process. Positives were made of each record, printing through *the base* of a special, thick film. This film, when developed carried a gelatine image on it in relief. Each frame could then function as a miniature printing plate. And it was this that gave Technicolor world leadership. The colour records were printed one after the other on to film that was blank, apart from the frame lines and soundtrack. The red record was soaked in a cyan dye, the green record in a magenta dye and the blue record in a yellow dye; each one being the primary subtractive colour. Precision alignment at every stage from camera to pin-belt printer, gave Technicolor images a brilliance and sharpness unmatched by any other system.

The first of the three-strip, beam-splitter, Technicolor cameras was completed in 1932. They were then put to work, first for Disney and then on the closing sequences of Eddie Cantor's *Kid Millions*, their first live-action use. The first full-length, dramatic film in the new three-colour method was *La Cucaracha* of 1934. The first feature Technicolor film followed in 1935. This was Mamoulian's *Becky Sharpe* which starred Cedric Hardwicke and Miriam Hopkins. In Britain the first Technicolor feature starred a racehorse *Wings of the Morning*, and was first seen at the Haymarket's Gaumont Theatre in London, in May 1937.

The Technicolor supremacy seemed invulnerable. Many rival systems were attempted but not one of them did more than produce a short or two, or some low budget movies best forgotten. Among the rivals were *Raycol*, first seen in 1928; the *Morgana Process*, first seen in 1933; *Rouxcolor*, first seen in 1932; *Splendicolor*, first worked in 1928; *Chimicolor,* first seen 1931; *Harmonicolor*, first shown 23 March 1936 at the Curzon, London (*Talking Hands*); *Harriscolor,* first used 1930; *Photocolor,* first used 1930; *Sennett Color,* first used 1930; *Multicolor,* first used 1931, and *Cinecolor*, a two-colour system first seen in 1930 and the only system to have a long life; it was in use for well over 25 years.

But the reign of Technicolor's bulky, costly, beam-splitter cameras came to an end with the perfection of monopack films after the Second World War. The monopack concept, three differently sensitive layers on one film, was first developed for cinema use in the early thirties. Karl Schinzel's *Katachromie* idea of 1905 was certainly the first *description* of a monopack but was never taken beyond the theoretical stage. So the man who made monopack a reality was Bela Gaspar, whose first important patent was issued in the US on 9 December 1931. By 1934 he was using his first colour film and had founded his first company, *Gasparcolor* Ltd, in Great Britain. He was at least a year ahead of *Kodachrome* and *Agfacolor*, the competing monopacks whose names are much better known. *Gasparcolor* was the first to offer commercial monopack prints;

the first to process on standard 35mm machines in a commercial laboratory and the first to carry a normal silver sound-track alongside a pure dye picture image.

The rival Kodachrome first arrived in 1935. It differed radically from Gasparcolor, since it *formed* dyes when developed, whereas Gasparcolor *destroyed* dyes. This makes Kodachrome the first integral tripack to use colour formers, the system we all now depend on for our colour photography. When first introduced it become popular as a 16mm reversal film, ideal for amateur use and occasionally for use by professionals in tight corners or difficult terrain (much of the *Conquest of Everest* was shot in 16mm). Kodachrome 35mm was then used to make the first feature film using nothing but monopack, though the release print was made by Technicolor printing. This was the 20th Century Fox production *Thunderhead – Son of Flicka* of 1945. But 35mm monopack had been first used just over two years earlier in 1942, during the making of MGM's *Lassie Come Home*, its use then, though, was confined to the outdoor sequences only. Kodachrome was never a full menace to Technicolor, that came with the perfection of another Kodak film, Eastman Color Negative, first introduced in 1951 under the name *Ektacolor*. In 1952 the first Eastman Color film *Royal Journey* reached the cinemas. Within the next three years Eastman Color had knocked Technicolor into second place. From then on Eastman Color became the only process considered by film makers, though it was often disguised by names such as *Warnercolor*, *Metrocolor*, and other house fancies. Technicolor survived only as a printing process, and a superb one at that.

Once colour was accepted as the norm other excitements were looked for. How could the cinema experience become more all-enveloping, more intoxicating? The logical solution seemed to lie in larger and larger screens to add the illusion of an extra dimension. The wide screen idea had first been tried out in a sequence of the Italian film *Il Sacco di Roma* of 1914. But the invention that gave wide-screen its commercial success was the special image-squeezing lens devised by Henri Chrétien in 1927. He called it the *Hypergonar*; we now know it as the anamorphic lens. It can take a wide picture and compress it sideways so that it fits a standard 35mm frame. The image is naturally distorted, but when projected through a lens that reverses the process, it can fit on to the wide-screen and look realistic. The first cinema system to employ Chrétien's lenses was Fox's *Cinemascope* of 1952. The old cinema screen proportions were 4 to 3. Fox used a new standard of 9 to 3 and the first feature in this new format was *The Robe*, premiered at the Roxy Theatre in New York City on 16 September 1953. Cinemascope offered one extra touch of escapism – for the first time it provided *four* sound-tracks giving the sense of movement to the sound.

Cinemascope came to Britain in 1954 when the first British Cinemascope production was shot. This was *Knights of the Round Table*. Thougha British film, its stars were Hollywood's Robert Taylor and Ava Gardner.

Yet even Cinemascope was not enough for some, and faced with the ever-present bogey of television, the industry began to cultivate an even greater wrap-round system and *Cinerama* was born. Wider still and wider stretched the screen and to fill its yawning yards three projectors were needed. And to cater for three projectors, three synchronized cameras were called for, matched and built together as one solid unit. Three matched taking-lenses of 27mm focal length, set at angles 48-degrees apart, covered a vast area and provided three filmed records that slightly overlapped at the edges. When projected on to the huge curved screen the effect was breathtaking. And seven sound-tracks using stereophonic techniques made the films ultra-spectacular. It was first shown to the public in New York on 30 September 1952, in a sampling type of programme called *This is Cinerama*. The first full-length feature had to wait for another ten years before it was made. This was *The Wonderful World of the Brothers Grimm* (MGM 1962).

Cinerama had its origins in a World War II training scheme for US air-gunners. In a more elaborate form (five projectors) it was first used in the Flexible Gunnery Trainer invented by Fred Waller. The trainee gunners would watch the realistic screened sky around them and learn to react fast to hostile aircraft as they weaved around and 'attacked'.

The Waller concept was taken to its peak by Disney when that company devised its *Circarama*. As first set up in Disneyland (1958) it featured a round screen, made up of eleven 8-foot by 11-foot panels. Eleven interlocked, synchronized projectors were used to fill the screen. The effect was said to be 'Particularly overpowering in the sense of motion, or moving with the picture.' One of the scenes was a 90-mile-an-hour auto ride!

Others looked to real three-dimensional films to fill their cinemas. The first attempts at stereo films dates back to 10 June 1915, when patrons of the Astor Theatre, New York were entertained by three films made on the anaglyphic system. This uses red and green images taken by camera lenses placed 62mm apart. Viewing was through spectacles with one red and one green glass. The results were forgettable. Even so the same system was later put to work to create the first 3-D feature film *Power of Love*, first shown at the Ambassador Hotel Theatre, New York, on 27 September 1922. The first feature-length 3-D film to break with the anaglyphic method was the Soviet film *Robinson Crusoe* of 1947. It was also the first feature length 3-D film in both sound and colour. It used what were termed 'radial raster stereoscreens' of grooved corrugated metal to deflect the two stereo images to the right and left eyes.

Yet another system was employed to make the first stereo feature with stereo sound. This used polarizing filters on the twin camera lenses to bias the images in different planes. Polaroid spectacles allowed right polarized light through to the right eye and left polarized light through to the left eye. As seen in the *Mystery of the Wax Museum*, in 1953 this gave a most realistic feeling of depth. Unfortunately polarization made the screened images darker than those of a conventional colour film.

For some, holographic film seems to offer the chance of 3-D films free from the handicap of special viewing spectacles. And in 1977, at the Cinematographers' conference in Moscow, the first holographic film was shown. The illusion of solidity was impressive, but since only a few people at a time could view the film, it is doubtful if the future lies with the holograph.

One thing is certain, the forecast threat from large screen television in cinemas is truly an illusion. The television image is at source broken up into sections. Any enlargement simply enlarges these sections, it can never eliminate them. Definition as a matter of course can never compete with the biting sharpness of photographic images projected through top-quality optical glass lenses. The first man to prove me wrong gets a medal.

SOUND • BROADCASTING •

Until the eighteenth century the only ways of communicating fast and reliably over long distances were limited by the speed of horses or runners. Other methods using drum signals or signal fires were too crude to be precise and dependable. A more sophisticated form of signalling was obviously called for, and there is a claim that this ideal was first reached way back in 1584. According to an Austrian chronicle, *Annales Ferdinadi*, John Dee the famous astrologer, alchemist, necromancer and geographer devised a so-called 'moonbeam telegraph' and sold the idea to the Hapsburg Emperor Rudolph II. This mysterious 'telegraph' was only put to use years later, on 29 March 1598, when it transmitted news of the defeat of the Turks at Györ (Western Hungary) back to Prague. Ten relay stations, placed at 25-mile intervals, were used to pass the messages. Exact details of the system have never been revealed, but it looks as if it was a simple heliograph method relying on reflecting light from large concave mirrors. In that case the most elementary code could stand for the words 'success' or 'failure'. Since there is no record of Dee's system ever being used again it must have been far too restricting for fluent messages.

Fluency only became possible with the *visual* telegraph; the first system able to speed complete words over long distances. In 1793 Claude Chappé set up a line of such telegraph stations between Paris and Lille, some 140 miles. The stations were wooden towers with movable beams fitted at their tops. Each tower was visible to the one in front of it; the beam positions indicated letters of the alphabet; and the positions were copied in turn by tower after tower. In that fashion the message was passed from Paris to Lille. And since every point in between had the message, this intelligence could be known to vast numbers, fast.

The first message signalled over Chappé's telegraph was sent on 15 August 1794. It came from Lille and told the government in Paris that the French army had retaken the town of Le Quesnoy. From then on the utility of the system was repeatedly demonstrated and it was extended, first to Strasbourg then to other parts of France. The British heard of these feats and Lord George Murray devised a rival method to suit the Admiralty. His first telegraph tower of 1794 used a signalling board pierced by large circular holes which could be closed by wooden shutters. The first chain of the towers was erected between London and Deal, in

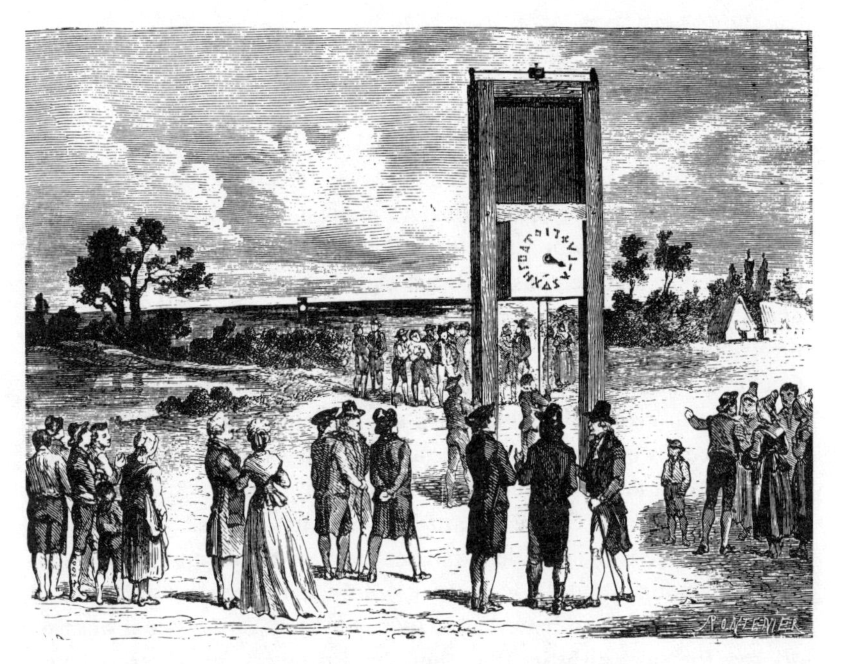

Chappé's first optical telegraph, 2 March 1791

Kent. Then a later chain extended to Plymouth, and over this Plymouth circuit a test message was sent to London and back at the incredible speed of three minutes.

The Prussians soon developed their own system, and in the USA a visual telegraph was first set up in 1800. This carried shipping news from Martha's Vineyard to Boston. So the value of fast news was appreciated within a few years; but there were drawbacks with the visuals. At night they were useless, as they were when fog or heavy rain struck. And each station had to be manned at all times. Thoughts then turned to an electrical method of transmitting coded words. The first electrical telegraph had been proposed by an anonymous writer in the *Scots Magazine* for 17 February 1753. His scheme involved one wire for each letter of the alphabet. At the receiving end each wire would have a ball at its end suspended just above a piece of paper carrying a letter of the alphabet. A frictional electric machine would be connected up to the right wire at the sending end; a charge would then flow through that wire and static electricity would attract the paper to the ball. This would be repeated until the whole message was sent. Now, our anonymous genius never even gave a clue to his age or nationality, but it is interesting to note that his ideas were first put into practical form by George Lesage of Geneva in 1774. So did Lesage write the letter of 1753, or did he simply 'borrow' the ideas?

No mystery surrounds later telegraph developments. In 1816 Sir Francis Ronalds erected a Telegraph system in his garden in London's Hammersmith. This used rotating dials engraved with letters of the alphabet at both ends of the circuit. These dials were clockwork driven and synchronized with each other. It worked! The Admiralty were informed of this success, but took no interest. They were quite happy to struggle along with their visual handicaps. In Munich S.T. von Soemmerring first demonstrated his electrochemical telegraph in 1809 and this inspired Baron Schilling to make his own telegraph. Schilling, in turn, influenced other experimenters, among them William Fothergill Cooke. Cooke was busy studying anatomy when he saw Schilling's apparatus in use in Heidelberg. He saw its faults and wrote 'I at once abandoned my anatomical pursuits and applied my whole energies to the invention of a practical Electric Telegraph'.

Back in England Cooke invented a synchronized disk system, along the lines of the Ronalds' telegraph, but it differed in an important essential. It abandoned the static electricity of the earlier system and put galvanic electricity (from batteries) in its place. The Cooke system was first offered for sale early in 1837, but the customer, the Liverpool and Manchester Railway, turned the offer down. They were wise to do so, since the problem of long-distance transmission had not been overcome. No one seemed to know how to grapple with the losses encountered with lengthy cables. And yet George Simon Ohm had supplied the answers in his book of 1827. This first explanation of the nature of electrical resistance remained little known for years, although its findings were crucial to the development of the electrical age.

Even when Cooke joined forces with Professor Charles Wheatstone the great snag of line resistance stayed with him. Long lines ate up power and gave such a small delivery that the electromagnets were not able to develop strength enough to function. Together they worked on a less ambitious system, one in which feeble currents could deflect compass needles and, by code, indicate which letter was meant. On 23 May 1837, Cooke made the first plans for laying telegraph cables under water. In the same month, by an odd coincidence, the Russian Government planned its first underwater telegraph line from Kronstadt to St Petersburg. Neither the British nor the Russian scheme was ever realized at that time. The chances are that poor line insulation would have defeated the plans.

More practical was the railway telegraph commissioned by the Great Western Railway. It was made to cover the distance between Paddington Station and West Drayton, some 13 miles west. This Cooke–Wheatstone telegraph first worked on 9 July 1839. It was then extended on to Slough in 1843 and this Paddington–Slough service made dramatic history when it became the first electrical system to catch a murderer. On 1 January 1845 John Tawell killed his mistress at Salt Hill near Slough. The cruel poisoning was soon discovered and, when making enquiries, the police heard of a man seen boarding a train at Slough. He was dressed in the noticeable garb of a Quaker and seemed respectable, but he answered the description of the man the police were after. There was no way of overtaking the train, so the police telegraphed his details to the telegraph clerk at Paddington. The clerk alerted the local police and a detective waited on the platform for Tawell's train to pull in. The killer was arrested, tried and hanged and the great publicity given to the case made millions aware of the telegraph and its speediness. For many years afterwards the lines from Slough to Paddington were referred to as 'the wires that hanged Tawell'.

In the same year, in the USA, Samuel Morse opened his first telegraph service from Washington to Baltimore. It first transmitted on 1 January 1845; its first message reading 'What hath God wrought?' This telegraph was the fruit of over ten years of thought and experiment. But though designated as 'The Morse System', it owed a great deal to an overlooked man, Alfred Vail.

Morse, a portrait painter with an electrical fixation, dreamed of messages by wire. His first demonstration of his infant system was made in September 1837, before an audience at New York University. This was a momentous event for among the onlookers was industrialist Stephen Vail. Vail agreed to help Morse with money and premises; in return he asked Morse to allow his son Alfred to work on the project. Morse agreed and Alfred, who was a telegraph enthusiast, transformed the clumsy devices made by Morse into simplified and robust apparatus. Morse had recorded his messages on paper bands, with a pen marking each letter in the form of a wavy pattern. Vail saw this as much too complicated and he constructed a code based on dots and dashes. This code, still universally used, is the famous 'Morse Code', but Morse had no hand in it, it was the brilliance of Vail alone that gave the world its first internationally accepted message code. Vail once wrote this: 'I am confident that Professor Morse will do me justice'; how wrong he was.

In Europe the first telegraph line was completed in France in 1845; in Belgium and Austria in 1846; in Italy in 1847; in Switzerland in 1852 and in Russia in 1853. Telegraph links were first forged between France and Germany in December 1850 but the ideal of international co-operation was given its first boost four months earlier when the first submarine cable was laid between Cape Southerland in England and Cap Gris-Nez in France. The date was 28 August 1850 and the first message seemed to herald years of cross-channel chatter, unfortunately a fishing vessel soon put paid to that dream. Its trawl raised the cable and it was chopped up as an interesting specimen of seaweed! A second cable, laid one year after, was armoured with iron wires and lasted for years before replacement. For the first time it was certain that submarine cables were worth large investments. And then the cables were slung between Wales and Ireland and Scotland and Ireland (1852); between England and Belgium and Denmark (1853); and between Italy, Corsica and Sardinia (1854). Outside of Europe cables were laid from Ceylon to India (1857); from Tasmania to mainland Australia (1859); and by 1860 London could telegraph direct to the Indian Continent. But the great Atlantic still posed the great problem. Special cables had to be constructed before the attempt to span it could start. The first Atlantic cables were devised by an American, Cyrus W. Field in 1857. On 7 August 1857 the first cable lengths were payed out at Valentia, on Ireland's west coast. Ten days later this cable broke and the voyage was abandoned. Work began again on 28 June 1858, but this time two ships, H.M.S. *Agamemnon* and U.S.N.S. *Niagara* met in mid-ocean, spliced their cables together and set off in opposite directions to their home ports. The cable snapped after an indecently short time. Teeth were gritted and the two ships met once more on 28 July 1858, and this time their efforts were not in vain. Both anchored on the 5 August and at 2.45 a.m. the first message passed across the Atlantic. It was prosaic and exact. It simply reported that the cable was open for commerce.

H.M.S. *Agamemnon* laying the first Atlantic cable, despite the intervention of a curious whale

On 14 August Queen Victoria sent the first Royal message over the cable; it was one of congratulation directed to US President James Buchanan. 400 messages sped through this cable, then the euphoria vanished. An operator applied far too many batteries to the circuit and the cable was wrecked. The American Civil War then halted all further cable-laying. When the next attempt took place the cable snapped. Finally, on 23 July 1866 the *Great Eastern* spanned the ocean with an intact cable stretching from Valentia to Trinity Bay, Newfoundland. The first reliable cable ever. Its first message was carried on 27 July 1866.

The advent of the Telegraph raised new hopes. Perhaps there was a way to send the signals from one place to another with fewer wires, or perhaps none at all? Failing that, perhaps one could send the voice itself along the singing wires? In other words people were looking for the Telephone and the Wireless. Prominent among these seekers was Professor David E. Hughes.

Hughes was the inventor of the first truly perfected mechanisms for printing telegraphs (1855). Others had tried their hands at developing such printers, but no one matched Hughes' brilliance. Success for his printing telegraph made Hughes rich and gave him the freedom to experiment and research the fields that excited him. He left his post in the USA, travelled throughout Europe and set up a home in London. It became the base for some remarkable discoveries, but the first step towards these discoveries was taken in St Petersburg in 1865 when Hughes was shown a strange device that sent musical sounds through telegraph wires. Even speech came through, though imperfectly. This device was in fact a telephone.

Now the Italian-American Society claims that Antonio Meucci invented the telephone in 1871. It was based on his experiments ranging over twenty-two years. The Italians themselves, on the other hand, insist that Innocenzo Manzetti of Aosta deserves the credit. And his monument at Aosta carries the inscription: 'Innocenzo Manzetti, inventor and maker of the first telephonic apparatus, in the year 1864'. Both Meucci and Manzetti, though, are rather shadowy figures, and so are their inventions. But even if we knew more about their work, neither man could qualify as the creator of the first telephone. It is now certain that the first authentic instruments used to transmit speech and music *electrically* were constructed as early as 1860. Their birthplace was a schoolroom in the town of Friedrichsdorf in Southern Germany. The man who devised and made them was a twenty-six-year-old teacher, Johann Philipp Reis. Fittingly he called his invention *das Telephon*. It was his device that Hughes had tested in Russia.

Reis' Telephones were made in a number of different shapes and sizes, including one shaped like a large wooden ear, but they all worked on the same principle. The transmitters employed two metal contacts held lightly against each other, whether by springs or by gravity. One contact was cemented to a taut diaphragm of gold-beater's skin and was connected to one pole of an electric cell. The other contact was wired to the distant receiver, and that, in turn, was fitted with a return wire to lead the current back to the cell. In use, the diaphragm vibrated when sounds were directed at it, and the pressure between the two contacts was varied. This continually altered the resistance of the circuit and controlled the flow of current to the receiver. The receiver reacted with physical movements that corresponded with the strength or weakness of the flowing current, and a faint, but understandable version of the original message or music was given off.

To give credit to Reis in no way detracts from the later work by Alexander Graham Bell. The two men were working on quite different lines. Bell had a distinct commercial end in mind, but Reis' telephones were not capable of being exploited commercially. They were sensitive, laboratory models, easily abused or put out of order. That was bad enough but, in addition, they were often misapplied and misunderstood. One lecturer even imagined that they were only intended to transmit the Morse code musically. Yet there is conclusive evidence that the Reis telephones were, and still are, capable of transmitting speech.

Four other firsts should be credited to Reis. He transmitted the first stringed music electrically, when he played a violin solo over his telephone. On the same day a colleague played a cor anglais solo over the 'phone, making this the first woodwind transmission. After that his 'square-box' model became the first telephone to be commercially constructed and the first to be put on sale. But it was never put to public use, the only demand for his equipment came from research laboratories or lone experimenters. And in view of the temperamental nature of 'das Telephon' that is all that could be reasonably expected.

The first *commercially feasible telephone* had to wait for Bell. Like Reis he began by studying the ear and at one time he even experimented with a human ear taken from a corpse. His master patent was filed on 7 March 1876. It has been described as the 'most important patent in world history'. Not everyone agrees! On 25 June 1876, Bell's first demonstration models were shown at the Centennial Exhibition, Philadelphia, where they were brought into prominence only after the Emperor of Brazil had all but swooned over them.

Bell's famous early models doubled as both transmitters and receivers and they could even operate without batteries by using the magneto principle. Inside the handle of these wooden 'butterstamp' models was a long iron rod in a permanently-magnetized state. Inside the large top-end was a thin, circular steel disk, positioned above a bobbin wound with insulated wire. Each end of this wire ran to terminals which could be linked to the land lines.

It operated on very simple principles. When strong enough sounds made the metal disk vibrate this altered its position above the magnetized rod. As Faraday had discovered, an alteration in magnetic flux around a wire will induce a current in that wire. The disk produced such alterations and the wire coil became filled with a fluctuating current that flowed along the land lines to the distant receiver. The receiver was simply a replica of the transmitter, and as its coil altered the state of flux in the hand-set, this caused *its* metal disk to mirror the movements made at the transmitting end. The original sounds, with some loss in quality, were then recreated.

Bell's first production telephones worked without batteries, since they generated their own power; but this power was unsuited for long-distance working. Commercial expansion obviously demanded change and this resulted in the substitution of special carbon microphones as transmitters; and these had to depend on a DC low-voltage supply. Bell's own invention still survived, but only at the receiving end.

The loose-contact carbon microphone was David Hughes' great contribution, first to telephony, then to broadcasting and sound recording. Its invention came about during an attack of bronchitis, in November 1877. Hughes was then penned in at home. The swirling fogs outside were deadly for his lungs, so he overcame boredom by playing around with a Bell telephone and musing over the idea of making it work somehow without any connecting wires. He already knew, from Sir William Thompson, that a wire would pass less current when thinned out. Thinning it would increase its former resistance. So, suppose that *sounds* were able to stretch and compress a wire carrying a current; would that not cause a telephone to recreate the original sounds? Thinking along these lines he made several futile experiments and then he stretched his test wire too much and it snapped. At the moment of snapping he heard a distinct grating and a sharp click in his telephone. He reasoned that the wire became sensitive to sound when in a fractured state and on rejoining the wire he kept the ends just pressing against each other. This imperfect contact led to the first recognition of the microphonic principle, for the new set-up allowed him to hear very faint sounds. From there he arrived at the epoch-making carbon rod microphone. Pencils of carbon formed a loose contact between small cups fixed on a sounding-board. These cups were wired into the circuit of a telephone. Then, as the carbon responded to vibrations, so the current fluctuated and animated the telephone diaphragm. The very footsteps of a fly could be heard with such a simple device!

Hughes refused to patent his discoveries and gave his knowledge to the world, free of charge, on 8 May 1878. This first explanation and disclosure of the secrets of the microphone was first given to the Royal Society in London and first given to the wider public on 9 June 1878. Amusingly the Royal Society reacted with both fear and elation. The Duke of Argyll thought that this discovery could lead to Cabinet secrets being listened in to; while Doctor Lyon Playfair argued that such microphones should be placed in both Houses of Parliament and wired up to compressed air amplifiers of the aerophone pattern. Speeches could then be heard over some 6 square miles around the Houses. This is certainly the first proposal for the public 'broadcasting' of Parliamentary proceedings, just as Argyll's fears represent the first forecast of 'bugging' operations. In fact, though, the first electrical transmission of a Parliamentary debate had already been made months before that historic meeting of the Royal Society. It was a private affair though, staged by the *Daily News* on 22 January 1878. As the words from the Commons came over the phone to the newspaper's office, so compositors set up the type for a special rush edition. But this was just a passing novelty, over a century ahead of its time.

The Hughes loose-contact microphone became the basis of improved models made by Gower, Crossley, d'Arsonval and Clément Ader. In Ader's hands the microphone reached a high point when he increased the carbon pencils to twelve in number. Using these ultra-sensitive models Ader discovered the stereo effect in Autumn 1881. He used this discovery to enhance a series of land-line broadcasts from the Paris Opera House to the Exhibition Hall at the Palais de l'Industrie. To create the stereo illusion he placed twelve of his large microphones on the Opera House stage, just behind the footlights. Thus six transmitters covered the left spread of the stage, while the other six covered the right side. As the opera unfolded the microphones responded with more power as the singers drew near them and gave a more feeble signal as the singers moved away. The cables from each side of the stage were directed through separate left and right conduits, and laid underground through the Paris sewers to the distant listening rooms. In practice each microphone proved capable of energizing eight telephone receivers, so forty-eight listeners at a time were given two receivers, one for the left ear, the other for the right. The left earphone picked up the sounds from the left side of the stage, while the other handled the sounds from right of stage. This gave an uncanny sense of movement and depth to the music heard through the phones, an experience now familiar to millions. These first stereo transmissions also qualify as the first public broadcast entertainment, for broadcasting in its fullest sense involves the organized electrical passage of sounds, speech and music from one place to another. This

An announcer with the world's first broadcasting station, 1893

makes broadcasting proper, far older than *radio*. Indeed the world's first broadcasting station was created many years before the first radio station became possible.

This first broadcasting station worked on a land-line system. It was conceived by the Hungarian electrician Theodore Puskas, a one-time collaborator with Edison. It opened up in Buda-Pesth in June 1893 under the name *Telefon Hirmondo* (Telephonic Newsteller). Each day, from eight in the morning to eleven at night Telefon Hirmondo sent programmes over its wires to the homes and offices of its thousands of subscribers. They paid very little for the service – something like one penny a day plus an initial charge for the special telephone receiver. And what did they get for their money? A typical daily programme shows that they were given time-signals; news bulletins from Vienna; foreign news reports; digests of the official and local daily newspapers; Exchange quotations and other financial reports; theatrical and sporting news and views; Parliamentary, Court, political, military and legal news; weather reports and fashion news. And that list only embraces the main transmissions up until 4.30 p.m. Then regimental bands took over and broadcast music for the next two hours. At 7.0 p.m. came a major shift when the station opened its lines to the Opera House. Listeners then had the chance to hear complete operas from start to finish, including the encores! Once a week special programmes were put on to entertain and enlighten children, while on Sundays a selection of church preachers could be heard direct from their pulpits.

Telefon Hirmondo also featured the first broadcast 'Commercials'. For a payment of two shillings, an advertiser could buy twelve seconds worth of time. He was then free to advertise anything from silk knickers to galvanized buckets, and his adverts were cunningly sandwiched in between news items, giving subscribers little chance to hang up.

So, to this station must go the honour of featuring the following firsts: Children's Programme; News Programme; Weather Report; Orchestral Programme; Religious Programme; and of course, Commercial Adverts. It was also the first to feature an Announcer, though the station itself termed him a Stentor.

There were attempts to imitate the Hungarian success in France and the USA. In France the '*Théâtrophone*' company offered land-line connections to theatres and to the Opéra and Opéra-Comique. In New York the Long-Distance Company broadcast quartettes and quintettes over its lines. Both the French and Americans made their first services available in 1893, but they never expanded them to match the comprehensive coverage offered in Buda-Pesth. In Britain the first land-line broadcasts were started in 1894 by the London '*Electrophone Company*', but this too never reached the Hungarian standards. It was more like the French '*Théâtrophone*' than anything else.

Until the advent of radio, there were no further developments in broadcasting of much note. The telephone service itself, though, was constantly being improved. In 1889 Almon B. Strowger began his work on an automatic method of routing telephone calls. By 1896 he was able to introduce the world's first automatic exchange equipment, making it possible for subscribers to dial each other direct, and bypass the much slower hand-connections made through switch-board operators. In 1900 Professor Michael I. Pupin of Columbia University, patented his method of loading telephone circuits with inductant coils. These developments made efficient long-distance telephony a reality. But wireless telephony still remained elusive.

Wireless telegraphy, by contrast, had made its first feeble steps as early as 1879. Once again it was David Hughes who made the first practical discoveries. In experimenting with his microphone and an interrupted circuit he found that he could hear sounds in a telephone ear-piece even when all connecting wires were removed. He made a crude portable receiver, left his clockwork driven circuit-interrupter in his rooms, and sallied out into Great Portland Street, London. As he walked up and down the street he noted that he could still hear the regular interruptions taking place at his house. These were still audible at a distance of 500 yards. More experiments led him to believe that he had discovered new, invisible waves. He called them 'aerial transmissions'; they were, in fact, the first recognized wireless signals. Unhappily his work came to an end when in February 1880 he failed to satisfy a group of visitors from the Royal Society. They witnessed his demonstrations but dismissed his reasoning, arguing that everything could be explained by what was known about induction. Afterwards Hughes wrote, 'I was so discouraged at being unable to convince them . . . I refused to write a paper on the subject'. But the conservative academics were prepared to revise their ideas when Heinrich Hertz presented his more thorough proofs in 1887. They were the first accepted demonstrations.

Hertz showed that electric waves could be sent across space with the speed of light. His transmitter emitted a spark 5cm long and his receiver, a wire loop with a small gap, responded by sparking across its gap at the same time. No wires joined the two pieces of apparatus, therefore electromagnetic waves had passed between them, without doubt. Sir William Crookes was quick to predict that these waves could be used for wireless telegraphy, but Hertz himself was dubious.

In 1894, two years after Crookes' wise prediction, Guglielmo Marconi began his first attempts to increase the range of the 'Hertzian waves'. He introduced a grounded antenna into the circuit of his transmitter and found that he could then pick up signals even when the receiver was sited behind a hill. At the same time Sir Oliver Lodge gave his first lecture on the phenomenon of resonance or tuning when using electromagnetic waves. It was the first exposition of this vital piece of wireless theory. He followed this by patenting (1898) the first adjustable tuning coils for use in both wireless transmitters and receivers.

In Russia Alexander Popoff was inspired by Lodge's 1894 lecture and began developing his own wireless apparatus. On 12 March 1896 before a meeting of the Russian Physico-Chemical Society, he sent in Morse the words 'Heinrich Hertz', and this message was instantly received and understood at the completely detached receiver. For this feat he is heralded by the Russians as the inventor of radio. The claim is not universally accepted.

In Marconi's case there are no doubts about his first triumphs in September 1895. And in February 1896 he arrived in England prepared to meet any tests. He took out his first wireless patent on 2 June; demonstrated his system to Sir William Preece, Chief Engineer of the Post Office, and on 12 December gave the first public demonstration of wireless at the Toynbee Hall in London.

Before his first public demonstration took place Marconi had been introduced to someone who looked like a deadly rival. At the War Office he met Captain Henry Jackson of the Royal Navy, and Jackson enthusiastically described how he was *already* using wireless telegraphy to signal from one end of his ship to another. In a short time he would be able to transmit from one ship to another. This dismayed Marconi, until he learned that Jackson was only interested in the military uses of wireless. He had no commercial plans for his inventions. So it is to Jackson that we have to allot the first wireless transmissions aboard ships and between ships.

Developments from then on were swift. The first permanent wireless station was opened in November 1897 at Alum Bay in the Isle of Wight. It was run by Marconi's Wireless Telegraph & Signal Company. The company's first customer was Lloyds of London, who bought equipment for their lighthouse at Rathlin Island and for their shore base at Ballycastle. The first messages from Rathlin Island were sent out on 26 August 1898. Marconi's equipment was next taken to the USA in September 1899 and was first used to report the results of the International Yacht Races back to New York City. Two months later saw the first installation of a wireless station aboard the US liner *St Paul*.

From then on shipping lines gradually came to see wireless telegraphy as a necessity, not a luxury. Proofs of its value came fast. On 28 April 1899 the British East Goodwin Lightship had sent out the first distress signal, following a collision with the steamship *R.F.Matthews*. As the new century opened many distress signals were sent and acted on but the first wireless-guided rescue to stir the world came on 23 January 1909, when the 15,000-ton *Republic* hit the Italian steamer *Florida* off the East Coast of the USA. Both ships were wrapped in dense fog. Signal flares were useless in such conditions so the ships were invisible to any would-be rescuers. But the *Republic* had its radio intact, and its distress signals were picked up by a shore base which alerted all shipping in the area. Within 30 minutes, the *Baltic* came to the rescue. Guided only by radio signals, the *Baltic* was able to manoeuvre through the blackness and save all the 1,700 people from both vessels. Without wireless no one would have even known of the disaster.

Eighteen months later wireless came into international fame once more. In July 1910 the wife-murderer Doctor Crippen fled from London aboard the Canadian Pacific's liner *Montrose*. With him he took his mistress disguised as a boy. The captain had his doubts about the couple and wired these doubts back to his company's offices. Back came a description of the couple hunted by Scotland Yard. The captain then confirmed that they were on board, and Chief Inspector Drew set out at once for Canada on the much faster *Laurentic*. The drama of the chase, the arrest, the trial and the hanging of Crippen convinced any doubters of the immense value of radio as a communicator. This was the first identification of a criminal by use of radio messages, and the first wireless-directed arrest. But, whether on ship or on shore all wireless messages until 1900 were restricted to telegraph form.

The breakthrough to speech came in December 1900. At that time the Canadian scientist Professor Reginald A. Fessenden, first spoke over the airwaves. This first wireless-borne speech was made over the short distance of one mile at Cob Point, Maryland. Fessenden then developed his system to give it greater range, and on 24 December 1906 he gave the first-ever Radio broadcast; an event heard five miles out at sea by ship's wireless operators. This was made from the station at Brant Rock, Massachusetts, and included a violin solo, songs, verses and a gramophone record of Handel's *Largo*.

Within a year the De Forest Radio Telephone Company was sending out regular broadcasts from its studio in the Parker Building, New York. These were experimental; in the main they consisted of gramophone record recitals, but after a while some live events were included and this resulted in the first broadcast talk, given by Harriet Stanton Black in 1909. Since she chose the subject of 'Votes For Women' this should also count as the first ever political broadcast. And to De Forest goes the credit for the first outside broadcast, when he transmitted from the

The Crippen sensation.
He was the first criminal to be caught by the use of wireless telegraphy

Metropolitan Opera House in January 1910. This event also provided Caruso with his first live broadcast. But it was a rival, Charles Herrold, who gave the most variety to those early radio fans. Broadcasting from his School of Radio in San José, California, he sent out a mixture of records, live vocals and the first weekly news round-up. He first began transmitting in January 1909 and by 1911 was the first to broadcast on a daily scheduled basis. And it was from the USA that the first transatlantic speech broadcast was made on 21 October 1915. It took the form of a message from the US Naval Base at Arlington, Virginia, to a representative of the French Government, waiting in a radio cabin built inside the Eiffel Tower.

By contrast, Britain lagged behind. It was the virtual first home of wireless. It had witnessed the first transmission across the Atlantic in 1901 (12 December). It had sent its first speech across the Atlantic on 19 March 1919. But for all, home listeners only heard their first broadcasts of speech in January 1920. These were scrappy programmes put out by the Marconi Company from its Chelmsford factory. (Occasional speech fragments accidentally picked up in earlier years, from 1913 onwards, were simply equipment tests made from Marconi bases.)

On 23 February 1920 the Chelmsford station began the first regular transmissions of twice-daily programmes, containing news, and music. The first male singer to broadcast on these programmes was tenor Edward Cooper; he was followed by the first woman vocalist to broadcast, soprano Winifred Sayer. Both these singers were amateurs, though Miss Sayer was paid five shillings, which could make her the first paid radio artiste. The first announcer on this venture was an amateur as well – the good-humoured Lieutenant-Commander W. T. Ditcham, of Marconi's Research Department.

These broadcasts caused a great deal of excitement and the *Daily Mail* focused on this and contracted Dame Nellie Melba to sing on the air. As a result, on 15 June 1920 listeners heard their first professional singer when Dame Nellie first sang *Home Sweet Home* then followed it with *Nymphes et Sylvains*, *Addio* from *La Bohème*, and *Chant Vénitien*. Considerably more than five shillings changed hands this time; Melba walked away with £1,000. The first large radio fee ever paid.

Melba's broadcast was such a triumph that the Danish operatic star Laurenz Melchior was next engaged to sing on air. His radio debut in that summer of 1920 makes him the first male professional singer to broadcast from Great Britain. He was also the last to broadcast for some time in Britain since the station was suddenly closed down by a Government order. The reason given was that Chelmsford's beamings were a hazard to communications with aircraft and shipping.

No such problems faced the broadcasters in Germany. The German Government had recognized the potential of wireless from the very start. When Marconi demonstrated at Salisbury Plain, there among the watchers was Professor Slaby of Berlin, sent to keep an eye on things and report back to the Kaiser. At that time the Kaiser dreamed of a German Colonial Empire and that meant a strong navy, as up-to-date as the British fleet. Since the British Navy were planning to use wireless, then the German Navy must have it too.

To make this possible government money was used to back the research projects of Slaby and his partner Count Arco. From this alliance grew the great firm of Telefunken, whose first aim was to counter the British-backed Marconi developments. One of the puzzling steps taken by Telefunken involved the erection of the giant transmitter at Nauen, near Berlin. This was so powerful that it could be heard in most parts of the world. It was the first station with such a mammoth range. But it was also the most expensive ever built. No amount of commercial traffic could justify its cost, so why was it there? The answer came on 2 August 1914 when Nauen began hammering out the enigmatic message, 'A Son Is Born'. Over and over it went out and every German ship at sea headed for home or the nearest safe port. It was a signal warning that war was due to break out. With that *one* message Nauen had paid for itself and saved the German Merchant Fleet. It was the first time that wireless had been used to control events on a world-wide scale. And the first major use of wireless as a military auxiliary.

Telefunken followed this triumph by controlling the movements of U-Boats in Atlantic waters. At Tuckerton and Sayville in the USA Telefunken had opened commercial stations operating on long waves only. Such long waves were useless for transmitting direct to U-Boats so the two stations sent Allied shipping information over to Nauen, and from there it was redirected on shorter wavelengths, matched to the sets on the submarines. This was the first use of wireless to direct attackers to targets.

When the war ended German radio was well advanced and needed no publicity campaigns to convince people of its great potential. After the Peace Treaties of 1919 the momentum was regained once more and on 22 December 1920, the Königswusterhausen Radio Station broadcast its first instrumental concert. From then on German radio technology became renowned, and justly so.

In the USA radio developed free from war-time restrictions until 1917, then it stagnated for two years. With peace its fortunes revived and in August 1920 the first commercial station with a regular daily schedule opened up. This was WWJ Detroit, which began earlier that year as 8MK, an experimental unit working out of the *Detroit News* offices. But the first US broadcasting company to skip the experimental stage and start off as a fully-fledged entertainment station, was KDKA, of Pittsburgh. This went on air on 2 November 1920 and its first broadcast gave the results of the Harding-Cox election. Within two years US radio acquired the seal of respectability when President Warren G. Harding made the first Presidential broadcast. The occasion was the dedication of the Lincoln Memorial in Washington DC, on 30 May 1922. And in 1922 came another important move, more important than any Presidential vaporizings. The US Courts decided that the American patent held by the English Professor John Ambrose Fleming, one describing his diode valve, was invalid. This meant that De Forest could now make and sell his three-electrode radio valves without facing legal actions. Both men had created the first valves of their kind. Fleming had invented the rectifying vacuum tube. But De Forest had gone further and invented a tube that could *amplify* incoming signals. As well as that his triode could be used to generate the high-frequency oscillations needed for radio transmissions. These valves now revolutionized the design and efficiency of all forms of radio apparatus.

In 1922 the US housed sixty broadcasting stations. Within a year the number had jumped to 588. In Britain, by stark contrast, there were four stations only; the tiny 2MT at Writtle, owned by Marconi; and three other stations destined to be incorporated in a somewhat larger body known as the British Broadcasting Company. This first BBC was a private enterprise venture set up by a group of six manufacturers of receiving sets. But it was from Writtle, not the BBC, that listeners heard

The first triode valve.
Lee De Forest's crucial device

their first broadcast play on 17 October 1922. This was a shortened version of *Cyrano de Bergerac*.

Before being swallowed up, the main London station, 2LO, gave the first Royal broadcast when the Prince of Wales made an address to the Boy Scout movement. This followed an earlier afternoon event at Alexandra Palace, London, where the Prince spoke to 50,000 scouts with the aid of amplifiers and public address loud-speakers. This was the first time that a Royal speech had been heard by a multitude. Then, on that evening of 7 October 1922, the gist of the afternoon speech was broadcast from York House, St James's Palace, as the high spot of the 2LO service.

The first programme broadcast by the new BBC on 14 November 1922 was mainly taken up with the polling results of the General Election. A weather report was thrown in for good measure. Entertainment came two days later, when vocal and instrumental pieces were transmitted. The first musical item heard was a baritone solo, *Drake Goes West*, sung by the BBC's first entertainer, Leonard Hawke. Comedy was provided first in November when Helena Millais assumed the character of 'Our Lizzie', the Cockney charlady she had first introduced on 20 October 1922 on 2LO. Programmes for children first went out on 15 November 1922. These were made at the regional station 2ZY at Manchester and called 'Kiddies Corner'. Religious broadcasts then began on the Sunday before Christmas 1922. This first helping of 'Radio Religion' was doled out by the Reverend John Mayo, Rector of Whitechapel, who was swept away with enthusiasm for the venture. He later wrote, '. . . I only ask this one thing of my executors – that they put on my tombstone: He preached the first broadcast sermon!' Christmas Eve also marked the broadcast of the first play written for radio. Written by Phyllis Twigg, it was a topical revelation of *The Truth about Father Christmas*.

Slightly more controversial was the first studio debate, between right-winger Sir Ernest Benn, and the Communist MP J. T. Walton Newbold, on 22 February 1923. Their topic, 'Communism would be a Danger to the Good of the People', was the first political discussion staged on air and marked the first contribution made by a Member of Parliament to broadcasting. Party political broadcasts on a small scale followed in 1924. When the October General Election campaign was in full sway, the Conservative leader Stanley Baldwin made the first speech for his party from the studios of 2LO. But the Labour Party leader made *his* speech from a public platform in Glasgow (13 October) and this led to the complaint that he '. . . did not broadcast well.' Ramsay MacDonald had overlooked the fact that the fixed microphone could not cope with his platform movements and oratorical devices of raising and lowering his voice. His is the first outside party political broadcast. Baldwin's counts as the first from a studio.

At this period, several new regular features were spawned by the BBC. The first of many broadcasts of opera was given live from Covent Garden on 8 January 1923. The whole of Act One of Mozart's *Magic Flute* was used. The first outside broadcast of a dance band (The Carlton House Dance Band) was heard on 28 May 1923. And the first symphony concert was broadcast on 16 June 1923. Film criticism had its first spot on 11 July 1923; drama criticism made its first appearance on 27 July and literary criticism began on a weekly basis on 3 September 1923. 6 October saw the first dramatized performance of a novel (Scott's *Rob Roy*); five days later saw the first complete Studio Opera, a performance of Verdi's *Il Trovatore*. Listeners could now read all about these treats since on 28 September 1923 the BBC issued the first number of its famous *Radio Times*, giving information and advance notice of each week's programmes. And on 31 December 1923 the warm chimes of Big Ben were broadcast for the first time. Radio made its chimes the first heard all over the world, and the most familiar of all time. Then in April 1924 King George V became the first monarch to broadcast when he opened the British Empire Exhibition at Wembley. He went on to become the first monarch to make regular broadcasts and gave the first of his renowned Christmas talks on 25 December 1932 during the first week of the new-born Empire Service (first heard on 19 December).

In its first form, as the British Broadcasting *Company* the BBC came to an end in 1926. It was replaced by the body that still exists, the British Broadcasting Corporation, which first took over on 1 January 1927, and became the first broadcasting concern to operate under a Royal Charter. This new BBC stayed at the original first home at 2 Savoy Hill until 14 May 1932, when all operations were switched over to the specially built 'Broadcasting House' in Portland Square. This new building was the first designed for broadcasting and nothing else. It was first brought into service on 2 May 1932 and its first broadcast after switch-over day, began at 10.30 on the morning of 15 May 1932. It was nothing more dramatic than a shipping forecast. But the night before, the BBC had staged a truly dramatic funeral for Savoy Hill. Its *End of Savoy Hill* was a great panorama of ten years of broadcasting in 161½ minutes. It was the first great broadcasting epic celebrating broadcasting itself. In that last programme the BBC reconstructed many of its early firsts, from talks to religious sermons, and even vaudeville acts. Many of these items were later placed on a disk for public sale; the first time anyone could buy a radio programme and play it as often as one liked.

All of the BBC's output was financed by the money raised from the licence fees; this scheme first came into force on 1 November 1922 when a Broadcasting Receiving Licence cost ten shillings each year. Since the broadcasts were independent of advertising revenue this gave the BBC a unique freedom from commercial pressures.

It was very different in the USA. Radio stations there were dependent on advertisers to buy air-time for commercials or even sponsor complete programmes. Commercial money became even more important when the many small stations began to group together into networks. The first of these big networks operated under the wing of the National Broadcasting Corporation and consolidated itself in 1926. The first of the big network shows was the *Ever Ready Hour* sponsored by the battery manufacturers. As a means of keeping stations on air, sponsorship seemed to be an ideal solution, but it had its black side. The sponsor could put pressure on the network to exclude items that in any way undermined their claims. They could, and did, object to certain scriptwriters or performers. In short they controlled with their reins of disapproval. Even top performers had to agree to plug products. Bing Crosby's first radio obligation was to promote the glories of Wrigley's Spearmint Gum. Bob Hope was saddled with the promotion of toothpaste or 'rack shellac', as Crosby jibed.

Though radio advertising was banned in Britain, business interests

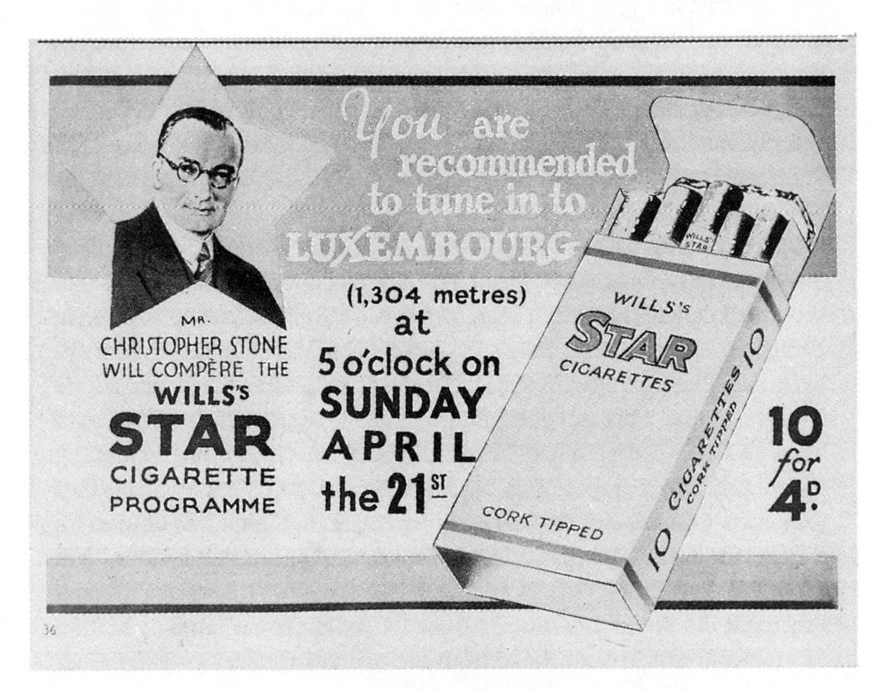

Commercial radio first invaded Britain from the Continent.
Ex-BBC stars helped improve ratings

soon found a way to feed adverts into British wireless sets. Early efforts in 1928 and 1929 first using Radio Hilversum, then Radio Toulouse were far from spectacular. But this changed in 1930 with the formation of the first dedicated, strictly commercial, English radio organization the 'International Broadcasting Company'. This was headed by Captain L. F. Plugge. In 1931 this IBC began its first commercial transmissions in English from the studios of Radio Normandie. These then led to further commercial programmes from the tiny Duchy of Luxembourg in 1934. And, until war intervened, the BBC began to feel threatened and even complained to the Ullswater Committee of 1935 about 'unfair competition'. Even after the war the BBC kept competition at bay and it was not until 8 October 1973 that the first commercial radio station (LBC) began to operate in London.

During the period of its monopoly the BBC became the first body to aim its broadcasts at the whole world. This began with its Empire Service in 1938 and was supplemented by the first of its foreign-language services (Arabic) on 3 January; the first Latin American services (Spanish and Portuguese) 15 March, and the first French, German and Italian services on 27 September. An Afrikaanse service opened on 14 May 1939, followed by European services in Spanish and Portuguese (4 June) and special English-language programmes for Europe in general on 1 August 1939.

When the Second World War broke out a host of new foreign-language programmes came into being. Services in Hungarian began on 5 September 1939 closely followed by Polish (7 September); Czechoslovak (8 September); Romanian and Yugoslav (15 September); Greek (30 September) and Turkish (20 November). On 7 January 1940 the first Forces Programmes began and on the same day the Bulgarian service opened. Twelve more foreign-language services were set up in 1940: Swedish (12 February); Finnish (18 March); Danish and Norwegian (9 April); Dutch (11 April); Hindustani (11 May); Maltese (10 August); Burmese (2 September); Belgian (both French and Flemish, 28 September); Albanian (13 November); Icelandic (1 December); and Persian on 28 December. Foreign-language expansion continued in 1941 with Slovene (22 April); Thai (27 April); Malay (2 May); Tamil (3 May); Cantonese and Kuoyo (19 May) and Bengali (11 October). Further expansion took place in 1942 with first services in Sinhalese (10 March); and in 1943 with Austrian (29 March); Luxembourgish (29 May) and Japanese on 4 July. The service in Russian was a late starter, its first broadcast did not go out until 24 March 1946 but it was more a creature of the 'Cold War' than a wave from the hand of friendship. (An earlier attempt in 1942 was dropped after a few months.)

For home consumption the BBC introduced its Light Programme on

29 July 1945 and its heavyweight Third Programme on 29 September 1946. The Third was the first programme to open with a tuning note. An A at 440 cps was broadcast each day, presumably for the benefit of orchestras and as a comfort for the hard-worked oboists, who normally gave the tuning A. Just as noteworthy was the first item broadcast on the Third; a first performance of Benjamin Britten's *Festival Overture*.

For many years all these domestic services were delivered on medium or long waves, then, on 2 May 1955 the BBC began its first VHF broadcasts from its new transmitter at Wrotham. By 1958 more than a dozen additional VHF stations were brought into service and on 13 January 1958 the first VHF stereophonic test transmissions were made from the London transmitters. Nine years later saw the introduction of Radio 1 (30 September 1967) and the first local radio experiment at Leicester on 8 November. But local radio had to wait for another three years before it was deemed acceptable. Then, on 4 September 1970 the first of the local stations opened at Bristol. Manchester's first station opened on 10 September; London's on 6 October; Oxford's on 29 October; Birmingham's on 9 November; Medway's on 18 December; Solent's on 31 December, along with Teesside's. Newcastle, Blackburn, Humberside and Derby had to wait until 1971 for their first broadcasts (2 January; 26 January; 25 February; 29 April, respectively). Within a few years all the local BBC stations were in direct competition with an array of commercial radio stations. Whether this was good for standards or not, is still debated, and ever will be.

• TELEVISION •

At an early stage television research involved two opposing ways of thinking. One camp pinned its faith on mechanical scanning systems; the other opted for electronic methods. The mechanical system was responsible for the earliest triumphs, for the first successful transmissions showing a living human face which were made by John Logie Baird using perforated disks revolved by electric motors. This was at an upper room in Frith Street, Soho, London, (a plaque marks the site) on 27 January 1926.

Baird, like C.F. Jenkins in the USA, had earlier only transmitted still pictures and silhouettes, but his continual refinements convinced him that his system was bound to win. He even showed that he could record and store television images on gramophone disks. This development of 1928 (*Phonovision*) must count as the first video recording system. His mechanically dependent flying-spot transmitter (it used a revolving mirror) was next used for the first transatlantic transmission from London to New York on 9 February 1928. Six months later Baird first demonstrated full colour television and then showed the first examples of stereoscopic television. He stubbornly resisted the argument that only an electronic system could give reliability and high-definition. He was strengthened in his beliefs by continental developments which depended heavily on mechanical devices, like mirror-screws and mirror-drums, as well as scanning-disks. He found an unlikely ally in the electronics pioneer Doctor Lee De Forest who wrote: 'Improvements in television will not come by way of the cathode-ray tube. They will come by way of mechanical reproduction. I know, I travelled all over Europe last year and saw all their equipment, but I am still convinced that the cathode-ray tube does not hold the solution to the problem, that it lies in mechanical reproduction.'

At the time De Forest gave that opinion (1935) cathode-ray tubes were delivering pictures at the rate of 24 per second, resolved at 240 lines. By contrast the Baird system was stuck at 12.5 pictures per second, with a crude definition of 30 lines!

The all-electronic system that we now enjoy was first detailed by A. Campbell Swinton, a British electrical engineer, in 1908. The system conceived a year earlier by the Russian, Boris Rosing, still stuck with the idea of mechanical scanning at the transmitter, even though his receiver was electronic. But it was one of Rosing's pupils, Vladimir Zworykin, who gave the death blow to all mechanical devices. Working in the USA he developed the first iconoscope camera tube, filing his first patent in 1923 and demonstrating his first working tube at a New York meeting of the Institute of Radio Engineers in 1929.

But, despite its excellence, the cathode-ray tube failed to conquer overnight and the first, regular, daily television service in the world was opened by US stations WRNY and W2XAL on 21 August 1928, using a strictly mechanical system. The first printed programme listing these television transmissions was carried by the *New York Times* on that day and shows a scheme mixing straight radio output with picture outputs at nine short periods each day. Prior to this WRNY venture the AT&T station in Whippany, New Jersey made headlines when it broadcast a short act by comedian A. Dolan on 7 April 1927. This makes him the first televised entertainer.

Even in those early days advertisers pressed the infant WRNY to accept its first commercials. *Radio News* of New York reported, in its November 1928 issue, that WRNY had been approached by a department store anxious to have its fashion show screened; by an advertising agency wanting air-time to promote a well-known brand of cigarettes; and by a large publishing firm wanting to have the cover of its latest book shown. These moves represent the first recognition of the selling power of television images, even such crude ones. Surprisingly, in view of the low state of development, 1928 also saw the advent of the first commercially manufactured television receivers sold by the Daven Corporation of Summit Street, Newark, New Jersey, and first advertised in July. This was before WRNY even went on air.

Yet, at the same time that WRNY and AT&T were rejoicing over their low-definition mechanical crudities, Philo T. Farnsworth was streamlining his electronic television system. He had first made a working system in September 1927, but it only sent and received simple shapes. By August 1930 he had secured his first patent for the first practical electronic system. It was a patent fiercely opposed by the Radio Corporation of America (RCA), but RCA lost the fight. Soon afterwards RCA's Vladimir Zworykin visited Farnsworth, and on seeing the first television picture said: 'Beautiful, I wish I had invented it myself.' This was the highest praise imaginable.

Yet the mechanical systems still led the field, and using GEC scanners, station WGY in Schenectady, New York, broadcast the world's first television drama on 11 September 1928. It was a two character play *The Queen's Messenger* by J. Hartley Manners and ran for 40 minutes. Not long afterwards came the first televised play in Britain. This was staged at the world's first specially created television studios at 133 Long Acre, London, home of the Baird company. It was *Box and Cox* transmitted on 15 December 1928 to an audience limited to the very few enthusiasts who had made their own sets from the plans published in *Television*. Any future expansion of Baird's audiences depended on an alliance with the powerful BBC. On his own he was hamstrung.

First reactions from the BBC were most unfavourable. Their chief engineer, Peter Eckersley, warned that the Baird apparatus had reached the limits of its development. It did not deserve a public trial. But strong publicity from the Baird camp led to political pressure on the BBC and it capitulated. On 30 September 1929 at 11 a.m., Baird made his first transmission from BBC station 2LO. It was a crazy event since he was only allotted one transmitter, which meant that he could broadcast sound or vision at any time, but not both together! By March 1930 he had won a two-transmitter session and he then made the first synchronized sound and vision broadcast in Britain. This was followed on 14 July 1930 by the first television play sent out by the BBC (though made by Baird & Co). It was Lance Sieveking's production of Pirandello's *The Man with a Flower in his Mouth*. This play introduced television make-up for the first time. Faces were painted yellow with features accentuated by thick blue lines.

Among the few people able to watch such advances was the then Prime Minister, Ramsay MacDonald. Baird had presented fellow Scotsman MacDonald with a receiver in March 1930; this was the first set ever installed at 10 Downing Street. And MacDonald was the first leader of stature to grasp the value of television. In awe he wrote to Baird, 'You have put something in my room which will never let me forget how strange is this world and how unknown'. Many have echoed these sentiments ever since.

MacDonald's set was one of the first factory-made receivers, almost everyone else was still dependent on home-made machines. Then, in May 1930 the first of the production sets were put on sale. For twenty-five guineas the public could now own an attractive *Baird Televisor* and dispense with their ugly wooden boxes and tangles of wires. Few took the bait. The price made the Televisor a luxury for the rich and when the Derby was first televised on 3 June 1931, the majority of viewers still watched in Heath Robinson fashion. It was a little different with the Derby of 1932. This time many lucky viewers were able to watch the finish on the world's first giant screen receiver. This measured 7 foot-high by 9 foot-wide and was erected in the Metropole Theatre in Victoria, London.

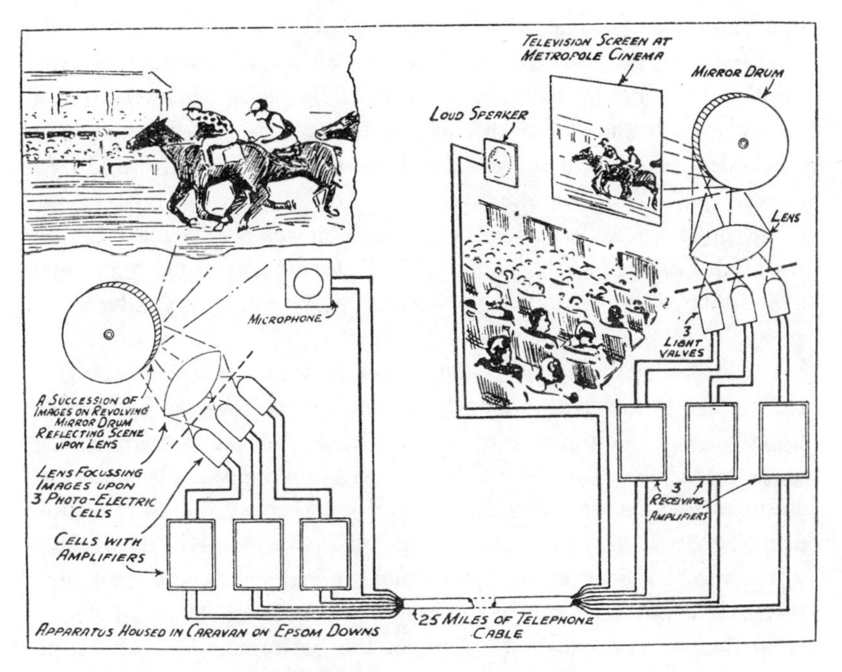

How the Derby was televised and shown to cinema viewers

All these advances were still hampered by low definition. Baird's company was still stuck at 30-line pictures, even though EMI had shown a mechanical scanner able to produce images at 180 lines. In 1933 Baird realized that he was being left behind and he first tried to catch up by adopting a German design for a 120-line film scanner. It was still mechanical but the results were way ahead of his own scanners. Unfortunately for him it was only of use in dealing with film. In a live-action studio it failed to cope. This failure led Baird to a temporary solution that gave him success for a while. He decided to film everything and develop the negative in a studio tank. As the still-wet film passed over the feed rollers it was then illuminated, scanned and transmitted at once. What the viewers saw as 'live action' was in fact real life delayed by the one minute needed to process the film. This ingenious 'Intermediate Film Technique' became the false saviour of the Baird company, since it gave it extra years but led it on to oblivion.

The first report by any government on television's future was that published on 4 January 1935. This was the British Government's Seldon Committee report, which recommended the opening of a BBC service based in a London studio. The committee was not able to decide in favour of the two systems that were competing to run the new service, so it compromised by allowing both Baird and EMI to run the service for a trial period. Each company was to broadcast on alternate weeks. EMI was quite happy to show the merits of their 405-line all-electronic system, but Baird was hard-pressed in his efforts to bring his scanners up to a higher level. In the end he did manage to reach the 240-line minimum demanded, but had to borrow a Farnsworth camera from the United States in order to stay in the race.

The first home of British national television was swiftly created at the Alexandra Palace, perched high in North London. And there, the first EMI studio was opened, while a separate studio was built for the Baird group.

The first regular service began earlier than planned, on 26 August 1936. This earlier date was an attempt to give life to the wilting Radiolympia exhibition at Earl's Court, London and was a snap decision made by the first Director of Television, Gerald Cock. The first programme (first from the new Baird studio) opened with an introduction from the BBC's first male announcer, Leslie Mitchell. His historic first words were: 'Good afternoon, ladies and gentlemen. It is with great pleasure that I introduce you to the magic of television'. Unfortunately a great deal of the magic was lost on the performers. Baird's set-up involved a cramped studio bedevilled by a flickering light (the 'flying spot' scanner); layers of the old heavy, crude make-up, and a rigid camera that could not be tilted in any direction.

The second day's transmissions were made from the EMI studios and their first-ever programme gave performers a new-found freedom and satisfaction. Make-up was simply as used in films. The studio was bright and roomy, and the camera was fully mobile.

The 'Radiolympia' affair was more of an emergency measure than a formal opening of the service. That came on 2 November 1936 when the BBC inaugurated the world's first public high-definition TV service. It was still lumbered with the Baird equipment which was used on each alternate week. Few people were happy with this fudging, and quite a number of artistes were reluctant to appear on 'a Baird Week'. The dilemma was resolved in February 1937 when the Baird system was discarded and for the first time BBC TV was placed on a modern, all-electronic basis.

German historians claim that Germany opened the first high-definition broadcasts twenty months before the BBC. But the system first used in Germany on 22 March 1935 was one using a 180-line picture generated by mechanical cameras; as such it fails to meet the high-definition standards which were set at an absolute minimum of 240-lines. To its credit, though, the German service did initiate the first coverage of the Olympic Games in August 1936.

In Britain the BBC made its own first spectacular outside broadcast when it covered the Coronation of King George VI on 12 May 1937. Cameras were not allowed in Westminster Abbey, but the imposing procession was filmed from platforms erected at Hyde Park Corner. Eight miles of thick cable stretched out from the cameras to distant Alexandra Palace and its transmitters. The King obliged by turning to look at the camera as his coach rolled by; thus becoming the first monarch to be televised.

Other European countries lagged behind. The French had first demonstrated 30-line pictures in 1929 and transmitted from the Eiffel Tower in 1935, but it did not reach its first high-definition standard until 1938, when it settled on 455-lines.

In the Soviet Union, the place where Boris Rosing had designed a cathode-ray receiver way back in 1907, the electronic camera had been worked on as early as 1931. But its first 240-line transmissions were not made until July 1938. And the first service did not begin until October 1938. This was Moscow-based and used 343-lines.

The United States also lagged behind, but part of the problem there involved complex court battles over the ownership of patents. There was no shortage of bright ideas or money to back them. In 1935, for example, RCA was prepared to lay out $1 million on its television demonstrations at New York's Radio City. Unlike Britain though, there was no talk of a single national service, everything had to be left up to private enterprise competition. And on 30 April 1939 NBC decided to open its first regular TV service. It chose the day well, for this was the opening day of the New York World Fair, due to be opened by President Franklin D. Roosevelt. NBC made history by telecasting the opening and Roosevelt became the first President to appear on television.

Even so NBC were only able to keep broadcasting for a year before the Federal Communications Commission stepped in to alter its status. NBC had to revert to an experimental role until agreement could be reached on national standards. At the time Philco was agitating for 605-lines as opposed to the 441-lines used by RCA-NBC. A compromise hit on the figure of 525-lines and the FCC decided that full-scale commercial television could now open up on 1 July 1941. The first day saw the first appearance of CBS as a newcomer to the big league and a direct competitor to the older NBC TV. It also saw the broadcast of the first television commercial, when the Bulova Watch company bought a ten-second slot to advertise its wares. This slot, on New York's WNBT, cost all of nine dollars! (An earlier first for Eugene Permanent Waves was never broadcast to receivers. It was a closed circuit item to be seen only at the Hairdressing Fair, at Olympia London, and was first screened on 5 November 1930.)

The Second World War closed down British television 'for the duration'. In most of Europe the adventure came to a full stop, and only German TV survived in a restricted form. In the USA war production halted the output of television sets and the service there entered into a state of limbo.

But at the end of the war the gears meshed once more and the war-weary countries began to look to television as the unexploited asset, for better or worse. In Britain the return to television was made at a snail's pace, and sets were still expensive, so this led to crowded living rooms wherever a set took root. Among the firsts viewed in those crowded rooms was the first post-war Olympic Games staged in London in 1948, and the Oxford and Cambridge Boat Race of 1949.

The great urge for set ownership came with the announcement that the Coronation of Queen Elizabeth would be covered *live* by the BBC cameras. It is claimed that more than one million people bought sets on the strength of this one Royal occasion, when for the first time five bulky television cameras were allowed into Westminster Abbey. Another fifteen cameras covered the route of the procession. This great event of 2 June 1953, also resulted in the first public demonstration of electronic colour television by the BBC. It was a closed circuit broadcast of the procession relayed to wards in the Hospital for Sick Children, in Great Ormond Street, London. (Earlier demonstrations of colour TV had all relied on mechanical scanning.)

This record-making first (some twenty million watched it), was the first sent live to France, Holland and West Germany. And tele-recordings were air-lifted to Canada and from there cabled to US television companies. In the USA some tasteless commercials were inserted into their screened versions and this lapse was used by those opposed to the growing lobby clamouring for a second, commercial channel. But the commercial lobby had power, influence and money. No surprise then when the Television Act of 1954 was passed, authorizing the first breaking of the BBC monopoly.

ITV, Britain's first commercial channel, first went on the air on 22 September 1954. Its first item featured a dinner at the Guildhall, in the City of London. As a symbol of the power and riches of the financial centre, this was most appropriate. The first commercial shown on that day was placed by Gibbs to advertise their SR toothpaste. The first jingle ran, 'It's tingling fresh. Its fresh as ice. It's Gibbs SR toothpaste'. For some though, commercialism still left a nasty taste in the mouth, especially when the end of *Hamlet* was faded out to make room for an orange squash advert. This was the first new ending since Shakespeare penned the work.

At this time both ITV and BBC were still broadcasting everything in black and white. The first colour demonstrations may have been British, but mainly due to the war, it was the Americans who first made colour television the norm. The first US service started on 1 June 1941 from the CBS station WCBW in New York. This never went beyond the experimental stage however, and so it was ten years later, on 25 June 1951, before the first regular colour broadcasts began. It was CBS again who made them and its first show was a sponsored variety presentation featuring Ed Sullivan. Its first series, *The World is Yours* presented by zoologist Ivan Sanderson, began on 26 June 1951. But the quality of the colour at that time was far from satisfying. This factor and the high price of sets gave colour a sluggish start.

Colour first came into regular use in Britain when the BBC opened its new service on 1 July 1967. One of the highlights of that day was the first colour coverage of lawn tennis at Wimbledon.

Independent television introduced colour just over two years afterwards on 15 November 1969; the first of its colourful commercials was a 30-second boost for Bird's Eye frozen peas. This great event (for Bird's Eye) hardly impressed viewers, for they were still intoxicated by the really significant TV spectacular of 1969, the 'giant leap for mankind', the first moonwalk by astronaut Neil Armstrong on 21 July. This has been dubbed 'The greatest show in television history', and for millions it had an odd, eerie fascination. Yet these were by no means the first pictures from 'out there'. Artificial satellites had already supplied these, when in 1962, satellite *Telstar 1* first intercepted, then reflected the first signals from the USA to Western Europe. These were limited to 18 minutes each orbit since *Telstar 1* was not high enough to reach what is now thought of as 'Clarke's Orbit'.

In 1945 Arthur C. Clarke first pointed out that a satellite orbiting at a height of 22,000 miles would revolve at the same speed as the Earth and would thus remain stationary over one point of the globe's surface. Radio messages could easily be bounced off such a satellite and redirected back. Just three satellites would give full radio coverage to the whole world. Clarke's dream came true in April 1965 when the US *Early Bird* (first commercial communications satellite) took up Clarke's 'stationary orbit' and provided 'live' shows from the US to Western Europe. These included the first great boxing contest, between Cassius Clay and Sonny Liston in May 1965 and international coverage of the 1968 Olympic Games from Mexico. By 1983 the European scene was enlivened by the launching of the first European Communications Satellite, this duly re-transmitted Sky Channel, Music Box and other programmes to all European networks.

The first television pictures beamed by satellite

Since then cable television has become married up to satellite transmissions giving a huge selection of programmes and an increased quality of pictures. Yet the cable idea is by no means new. It was first used back in 1927 by the Bell Telephone Company on an experimental basis, between New York and Washington DC. The idea was taken up with enthusiasm by the BBC's head of engineering Captain Eckersley, but never acted on. It was a small town in Oregon that in 1949, first benefited from the purity of image provided by cable. Surrounding mountains had previously made the pictures from Seattle jumbled and unwatchable. And all the first cable companies initially catered for areas where reception was poor or hopeless. But with time, even areas of good reception became targets for the cable system. Then in 1970, an old invention first patented in 1929 by Affel and Espensched (USA) was developed to an advanced stage. Optical fibre cables came in and gave television addicts such a huge selection of channels that it all now seems to be a twenty-four-hour-long feast with unlimited choice. And if the day is not long enough, then any video recorder will store up pictures for future display.

The video record in its first clumsy form dates back to Baird's *Phonovision* patented in 1927. But his disks, with the very little information on them, were not the way forward. The real way to advance only came after great improvements were made in the recording of *sound* on magnetic tape. Once this point was reached the logical next step was towards developing machines and tapes able to deal with the much more complex television signals.

The first demonstration of video-taping took place in 1951 but the first acceptable images and machine were not shown until 1954 when RCA displayed its first offerings. In Britain one year later, the BBC showed that it too had reached the goal. Its first Vision Electronic Recording Apparatus was certainly efficient, but at a price; since its ultra-fast speed led it to use up over *10 miles* of tape each hour! Fortunately the Ampex company had one eye on tape economy and it was their much more realistic VR 1000 that made the first retransmission of a programme possible. This took place on 30 November 1956 when the CBS studio in Hollywood recorded the 'Douglas Edwards and the News' programme from New York and then rebroadcast it, three hours later. At the same time the first magnetic videotapes were placed on sale. These were 2-inch wide Scotch tapes developed jointly by Mel Sater and Joe Mazzitello for their company 3M. This was the tape used by Ampex for its historic first retransmission. A mere two years on and Ampex was able to introduce the first colour video-recorder, their VR 1000 B. Ampex fittingly went on to provide a home for the world's first Museum of Magnetic Recording, presided over by expert Pete Hammar. In Germany BASF set up a similar tribute to the great first names in magnetic recording.

For some years video recording was strictly for the industry, until in 1964 the Japanese firm Sony marketed the first video cameras meant for home use. These were smaller versions of the types used in professional recording and used reel-to-reel tapes which had to be carefully threaded through guides, rollers and around the record/playback drum, where the heads were located. Few ended up in the home. Most went to schools and colleges who could justify the investment. It was not until October 1970 that the first easy-to-use video-cassette-recorders were first offered to the public; and here the European firm of Philips took the lead with its VCR. For a spell it remained unchallenged but in 1975 Japanese manufacturers Sony and JVC offered a challenge with their more compact systems. The first Sony machine used the Betamax system, a reduced scale version of the professional U-Matic system of high prestige. JVC used an incompatible format known as VHS on *its* first machines; so the companies who had joined hands earlier to create the U-Matic, were now in open conflict with each other.

The first system to die was the first offered, Philips VCR. But Philips later recovered momentum and offered a marvellous new system, the 200. This had superb freeze-frame, slow motion and long playing double track tapes. It should have triumphed but it came too late, for VHS had rapidly won the largest share of the market and Betamax was a good second. Then Betamax too went into decline, overcome by the more aggressive marketing of VHS. In the end the VHS system stood first without any rivals in the home. All fresh domestic developments were to be within its framework. These have included: the first reduced-size cassettes easily adapted for standard replay (1982), the first high-quality machine to increase the horizontal lines from 240 to 400 (March 1987), and the first *Window* recorder. This machine can produce special effects; stop at a given frame; permit viewing of tape and television at one and the same time; and allows the viewer to watch one channel while eight other small windows, showing programmes appear on screen. This dazzler arrived in 1987.

Video disks have trailed behind in the public's affections. They seemed full of promise but the first attempts were hardly encouraging. In 1965 Westinghouse made the first commercial probes with its *Phonovid* disks but they offered 200 still images only. The first marketing of a finer system (*Teldec*) created by Decca and AEG in 1975 was a flop. Philips first laser-read video disk (*Laservision*) of 1980 was a step in the right direction, but other makers went for different standards adding confusion to the sales problems. The first US system, for example, would only work on RCA equipment. RCA's *Selectavision* of March 1980 died in under four years. JVC of Japan went for a standard poised between RCA's and Philips' (VHD) and then replaced it with their better AHD system, though both these ventures were compatible. Then JVC's Japanese rivals decided to back the European Philips company in its development of Laservision and in 1984 Pioneer designed the first compatible reader of both CDs and Laservision and shortly afterwards made the first Laservision disks with digital sound. In 1986 JVC kept its end up by offering the first images in three-dimension. But nothing halted the Philips' inspired momentum, and in 1988 the peak of the momentum produced the first Video CD. With its backing by Philips, Sony and Pioneer this looks like the winning disk – at last.

• SOUND RECORDING •

In 1857 Edouard-Léon Scott de Martinville, of Paris, demonstrated the first machine to record sound. But he was not able to develop a play-back machine fit to reproduce the sound traces captured by him. Despite this his *Phonautograph* is of major importance since it led direct to the invention of the *Gramophone*. His prototype was beautifully simple. At the narrow end of a megaphone he stretched a tough membrane of gold-beater's skin. Fixed to the centre of the membrane was a light rod that ran parallel to it and rested on a pivot at the edge of the diaphragm. At right angles to the end of this rod he glued a stiff hog's bristle. This recording-head was fixed in front of a brass drum that could revolve and move sideways at the same time. Around this drum he fixed a sheet of paper evenly smoked in the flame of a wax taper. In use the megaphone was so adjusted that the bristle just lightly touched the sooty paper. When the drum turned the bristle left a thin, straight white line on the paper but when speech or music was directed into the horn the bristle traced wavy lines that were, in fact, the world's first sound-tracks. But they were only microscopic in depth and quite unable to guide a playback stylus. As a result the machine was only used by physicists who wished to study the nature of sound-waves using purely visual techniques. Twenty years later Scott's compatriot, Charles Cros, showed how these flat, undulating traces could be turned in to shallow grooves capable of being played back and listened to.

The first working machine able to both record and reproduce sounds was the brainchild of Thomas Edison. One of his telegraph experiments involved a sounder that could emboss its signals on a revolving and grooved paper disk. Thinking around this device inspired him to reason that a similar machine might be able to record the movements of a telephone diaphragm and in that way store up speech. In July 1877 he actually recorded and played back the first word ever, a crude, garbled 'Hello', indented on a band of waxed paper. Out of a string of rough experiments he then finalized a scheme for the first *Phonograph* and gave the specification to his fine mechanic John Kreusi, who completed the machine on 6 December 1877. It was a recording and reproducing machine in one, with separate heads for each function.

The first recordings were made on sheets of stout tin foil wrapped around a 4-inch diameter brass drum. This drum was cut with a thread of ten-to-the-inch and was mounted on a lead screw threaded to the same standard. A crank-handle at one end served to rotate the drum and advance it at the same time. The recording head was so adjusted that the stylus pressed the tin foil into the groove underneath it. When speech or music was directed against the diaphragm, the stylus moved in and out, indenting the soft foil with a hill-and-dale sound-track. It was an indentation method, pure and simple; no material was removed by the recording act. To play back, the recording head was disengaged, the drum was wound back to the starting point and the more sensitive reproducer was pushed forward until its stylus fitted into the grooved foil.

When the handle was cranked the stylus bobbed up and down, guided by the indentations and the playback diaphragm reproduced the original sounds, or rather some of the original sounds. The first words recorded and reproduced on this machine were 'Mary had a little lamb its fleece was white as snow . . .' As can be well imagined, this process was grossly insensitive and full of distortions. This was admitted by Edison in later years when he wrote '. . . it was a very imperfect machine and only produced a caricature of the human voice . . . no one but an expert could get anything intelligible back from it.' It was also a method that did not lend itself to mass duplication of recordings. Despite all these drawbacks Edison did make some remarkable forecasts of the future of sound recording. He was the first to recognize that when perfected, it would have many valuable uses. His first set of forecasts were published in the *North American Review* of June 1878; he said that it would usher in: '1. Letter writing and all kinds of dictation without the aid of a stenographer. 2. Phonographic books, which will speak to blind people without effort on their part. 3. The teaching of elocution. 4. Reproduction of music. 5. The "Family Record" – a registry of sayings, reminiscences, etc., by members of the family in their own voices, and of the last words of the dying. 6. Music-boxes and toys. 7. Clocks that should announce in articulate speech the time for going home, going to meals etc. 8. The preservation of languages by exact reproduction of the manner of pronouncing. 9. Educational purposes; such as preserving the explanations made by a teacher, so that the pupil can refer to them at any moment, and spelling or other lessons placed on the phonograph for convenience in committing to memory. 10. Connection with the telephone, so as to make that instrument an auxiliary in the transmission of permanent and invaluable records, instead of being the recipient of momentary and fleeting communications.'

These insights, of course, were glimpses of a new machine, for his first machine was sadly crippled by the wrong technology. Indeed, when it

The first talking dolls used tiny phonograph cylinders

was first introduced, the French inventor Charles Cros said of it: 'Mr Edison has been able to construct his machine. He is the first who has ever reproduced the human voice. He has accomplished something admirable.' Then he added, 'Despite the charming advantage of immediate repetition, despite the easiness and simplicity of its use, I do not think much of the future of the tinfoil Phonograph.' And here he was right on target. Edison's tin-foil indenting system did not lend itself to any type of fruitful development. The phonograph only took on a new lease of life when the sound-tracks were *engraved* instead of indented, and when the recording medium was changed to wax cylinders. In these vital developments Edison was to take second place to his rivals Alexander Bell and Sumner Tainter.

Bell and Tainter designed the first recording heads set to *engrave* a sound-track. They chose beeswax for their first records, but to speed things up, they did not bother to make a special machine. For their first demonstration they took a standard Edison tin-foil recorder and filled its deep grooves with wax. It was then fitted with their newly developed recording head with its cutting edged stylus. Their first trial machine was, therefore, a hybrid, but it worked on fresh principles and it gave the first engraved sound-tracks.

They deposited their first experimental hybrid, complete with its engraved recording, with the Smithsonian Institution on 20 October 1881. This first engraved sound-track carries the words: 'G-r-r, G-r-r- There are more things in heaven and earth Horatio than are dreamed of in our philosophy G-r-r I am a Graphophone and my mother was a Phonograph.' Also deposited with the Smithsonian was a sealed envelope containing hand-written copies of Bell and Tainter's first experiments and thoughts on new methods of sound recording. This was delivered over a year before the *Graphophone* was deposited. These first crucially important papers were given to the Smithsonian on 28 February 1880. They show, beyond doubt, that sound recording was liberated from stagnation by their Volta Laboratory. Edison had been content to leave his invention stuck at its original novelty stage, despite his original far-seeing words.

All recordings up until then had been made on tin-foil sheets which were only good for a few playbacks. That is why we no longer have the voice of US President Rutherford B. Hayes, first politician to record (White House, April 1878), or the first music piece recorded, cornetist Jules Levy's rendering of *Yankee Doodle* made mid-1878. 1878 also saw the first attempt to market the machines in the form of the $10 *Palor Speaking Phonograph*, and these were even advertised for the first time in the first copy of the *Boy's Own Paper*, on 18 January 1879. But few bought and as the novelty value declined, so did Edison's interest. He turned to more lucrative and interesting inventions, until on 27 June 1885, Bell and Tainter applied for their patent on wax-coated cylinders played with a flexible stylus mounted on a mica diaphragm. Edison met this challenge with his own wax recordings. His were *solid* cylinders, unlike the rivals, which were of cardboard under thin wax, a combination which made them susceptible to cracking. Bell and Tainter were wise enough to recognize that Edison's cylinders and machine were far superior to their products. So they ended the rivalry by negotiating a merger. The Bell machine, the Graphophone, first wax-cylinder machine ever manufactured commercially (1887), was allowed to die off – only the name was retained. Edison's *Perfected Phonograph* then set the standards and the first entertainment cylinders were made as hollow wax castings measuring 4-inches by 2.25-inches in diameter. When first marketed though, on 16 June 1888 it was first thought of as an office dictating machine, and fitted with two heads, one for recording on blank cylinders. And a special shaving device was provided to skim the cylinder's surface smooth after each recording.

The first Perfected Phonograph to reach London went to Edison's agent Colonel Gouraud on 26 June 1888. The Colonel then used it to make the first-ever recordings of famous contemporaries including Robert Browning, Florence Nightingale, Alfred Lord Tennyson and Gladstone. These were made in the first British recording studio created in his house, *Little Menlo* in Upper Norwood, London.

The first serious music recordings were made at Edison's laboratory in 1888 by the pianist Josef Hofmann, but these were private cylinders never intended for the public. In London, Colonel Gouraud made the world's first live (non-studio) recording when he took his machine to the Crystal Palace on 29 June 1888 to record part of Handel's *Israel in Egypt*. Yet another non-commercial event was the first performance recorded by a serious composer, an 1888 cylinder of Brahms playing one of his Hungarian Dances. But Edison's grandest recording-first was staged in 1889 when he attempted to record the whole of a concert conducted by Hans von Bulow at the Metropolitan Opera, New York. These first orchestral recordings, including the first of a complete symphony, were never issued and have vanished.

The very first commercial cylinders, made on 24 May 1889, seem at first glance, to be a matter of dispute. Some accounts state that a mixed batch of flute and piccolo solos were recorded by one Frank Goede. Other accounts state that fourteen flute solos were recorded by Edison's favourite flautist Eugene Rose. And Rose himself stated this in Edison's presence, without contradiction. This conundrum is easily resolved by referring to the practices of the day. At that time it was customary for entertainers and musicians under contract to use assumed names when taking on outside work – some musicians used up to six such names! And the term 'piccolo' was then not an exact term. It did not apply to one instrument and one only, on the contrary it could cover any one of five of the smaller members of the flute family. But big or little, to a *bandsman* they were all flutes.

The first recording violinist ever, made his cylinders one day after the flautist; this was Alfred Amrhein. On 28 May it was the turn of the first brass player when John Mittauer recorded polkas and army bugle calls on his cornet. The first duets and trios were then recorded on 29 May. Bassoonist John Helleberg provided a bass and the first rich bassoon notes on record; while Henry Giese became the first clarinetist to record and Max Franklin became the first pianist to record commercially.

All these first recordings were made one at a time, then on 3 June 1889 came the first multiple recording when flute solos were played into seven recording machines simultaneously. Singers were late-comers, since the first did not record until 1 June 1891. This was comedian George W. Johnson who belted away for two and a half hours. The female voice first cut wax on 9 July 1891 when a Miss Stewart sang *My Love and I* and *Le Pré aux Clercs*.

These early performers had to repeat their pieces over and over again to create a worthwhile batch of cylinders, but this way of working was crude and limiting, and a duplicating system was an urgent priority. Duplication first arrived on 18 March 1892 in the form of a dubbing machine which would copy up to 150 cylinders from one master. It was not the complete answer since the master's grooves became worn out by the process but it made for uniformity and cheapness.

The first record catalogue, issued by Columbia Phonograph Co. of Washington DC in 1891, lists 194 cylinders, all of them lightweight popular items. This same company marketed the first clockwork-powered cylinder-player, The Graphophone in 1894; though some add-on clockwork motors had been offered earlier by independent makers. At this point Columbia were ahead of Edison but all their basic standards, record dimensions, groove spacings were taken from Edison. Edison's own spring-motor machine appeared in April 1896 and in design eclipsed his rivals. His playback heads, for one, were far more responsive and robust. But while Edison worked at an efficient way of moulding cylinders he was overtaken by the much smaller Lambert Company of Chicago, who introduced the first moulded cylinders early in 1900. And they added a bonus by making these cylinders the first made from celluloid, a step which reduced wear to a minimum. Edison's first moulded cylinders appeared in 1901 but it was not until 1912 that he made the move over to tough, long-lasting celluloid. By then the phonograph was on its way out.

Before it succumbed to the disk machines, the cylinder interests had made some valiant attempts to offer advanced recordings. Large 5-inch diameter *Concert Cylinders* were first sold in December 1898; these were louder and more faithful to the original. And with the standard cylinders the grooves were increased from 100 to the inch to 200 to the inch; a harder wax made this possible, thus playing time was dramatically doubled from two minutes to four in October 1908, when these first Edison *Amberols* were sold.

By 1912 the cylinder era was in decline. Columbia had stopped production and Edison was developing his first disk records. But while they lasted the cylinders had brought speech to 'Talking Dolls' (1894); to Sivan of Geneva's 'Talking Watch' and his aggressive alarm clock (1895); and authentic speech to language teaching courses (Dr Rosenthal's

International Institute of Languages Course, 1893). Apart from that they had given multiple entertainment in coin-operated jukeboxes, first installed at the Palais Royal Saloon, San Francisco on 23 November 1889. And the first jukebox offering a pre-selection facility was cylinder-fitted as well. This *Multiphone* of 1905, held twenty-four numbered cylinders mounted on a huge wheel placed behind a glass panel. A crank put the wheel in motion until the selected number stood in the playing position, then music poured out of the 7-foot-high lyre-shaped wooden cabinet. All provided for a mere nickel.

The first teaching machine taught languages

On the debit side we learn that the cylinder became the first carrier of vocal obscenities. In May 1905 newspapers in New Orleans picked up the choice story that the phonograph was being besmirched in the interests of 'lewd and sordid activities'. In many of the city's brothels the machine had become renamed *The Pornograph*, because of its use to play grossly obscene ditties and verses. For a few extra cents the clients could even record their own foul utterances and hear them repeated while they took their pleasure. The Devil had truly found work for idle tongues! Yet the chances are that it was all so very tame compared with some of the vocals that, in time, were enshrined on the all-conquering disks.

THE GRAMOPHONE

The Gramophone (a trade name), or first disk record-player, is usually credited to Emile Berliner, but up-to-date research shows that at best Berliner was simply the developer who launched the disk-player on to the market in 1889. His role was crucial, but a study of his patent of 1887 shows that the essential ideas in that document were set down by Charles Cros in 1877–8. Indeed the very language used in the patent is strangely similar to that used by Cros nine years earlier. Cros' plan involved recording on a glass-disk coated with lampblack. His aim was to obtain an extremely delicate tracing, made by a light stylus grazing the coated surface. The stylus would move laterally, to reduce friction to the minimum, and the recording would be rendered as an undulating spiral line of clear glass, set against a dark background. This master record was meant to be used as if it was a photographic negative. A positive print made on a second glass could then be contact-printed on to a metal-disk coated with a gelatine-bichromate varnish.

This light-sensitive varnish hardened wherever light struck it and became insoluble. A rinse in water would wash away all the areas of unexposed, soft varnish and the sound-traces would be seen as a thin spiral of bare metal. It was then a simple matter to convert these thin, superficial lines into grooves deep enough to guide a play-back stylus. The answer lay in the photoengraving technique. The metal had only to be immersed in a bath of etching acids and the bare metal lines would be eaten into and given depth. (It was this technique exactly, that was used by Emile Berliner when he created his first disk records.)

But this was not Cros' sole idea. He also advanced the proposal that recordings could be made on wax-coated disks which could then be electro-plated and used as stampers to make many identical copies. Thus as early as September 1877 he described the duplicating system still used today in the making of standard gramophone records.

Cros was unable to construct his machines, since they needed a degree of precision that could be provided only by expensive lathes and skilled mechanics. The famous instrument maker Breguet was quite willing to make the prototypes, but his charges would have been 30,000 francs, in advance; a large sum, well beyond Cros' means. Today, the French Academy of Sciences hails Cros as the Inventor of the Talking Machine, but while he lived it was unwilling to contribute a single franc towards his researches, neither did it honour him in any way whatsoever!

Ironically enough, Emile Berliner was able to put Cros' ideas into practice with the aid of funds that originated with the French Academy of Sciences. The Volta Laboratories in the USA were given an award by the Academy and part of this money was paid to Berliner for use of his carbon-button transmitter patent in the Bell vs Edison battles. His epoch-making first patent for the *Gramophone* was first applied for on 26 September 1887. He first demonstrated a working machine at the Franklin Institute, Philadelphia, on 16 May 1888; yet his progress from then on was slow in the extreme. He was a plodder, rather than a dynamic figure – and far from an original thinker. It is worth noting that once his patents were safely past the rejection stage, Berliner belatedly acknowledged '. . . to Charles Cros belongs the honour of having first suggested the idea of and a feasible plan for mechanically reproducing speech once uttered.'

Berliner's first machines of 1889 were hand-cranked toys using 5-inch disks made of a hardened rubber composition. The first Gramophone record catalogue was issued in 1892 by Parkins and Gotto, the London importers of these toy machines. The recordings themselves were first known as 'plates'. Then, with the formation of the first Gramophone Company (US Gram. Co.) in 1894, followed quickly by the larger Berliner Gramophone Co. of October 1895, came a fresh evaluation of the adult market. This resulted in the emergence of an adult Gramophone late in 1896. It was driven by a DC electric motor. At the same time 7-inch, single-sided records pressed in vulcanite were issued. But electric-drive machines never caught on; the Gramophone only became popular with the introduction of the clockwork, spring-motor model of 1899.

The first Gramophone was simply a toy

To cater for this new market the first recording studio was set up in Philadelphia (on 12th Street) in 1897. The first recording manager was Fred Gaisberg, chosen because he knew many potential recording artists from his earlier days with pioneer phonograph companies. Finished recordings were sold from the first special retail record shop on Chestnut Street, Philadelphia. This opened in 1897. Its first manager was Alfred Clark and among his stock he offered the first dance records made by Sousa's band and the first ragtime recording *Cotton Blossoms*.

Sousa's band was organized along military lines, so the first disks cut by a *dance band* proper, were the polkas recorded by the Hotel Cecil Orchestra on 2 September 1898. These were all 7-inch disks meant to be played on the improved machines powered by the Eldridge Johnson clockwork motor, first demonstrated at the end of 1895. The Johnson machine shop, of Camden, New Jersey, became the first commercial manufacturers of the new Gramophone and its machines were shipped to Europe to coincide with the first recording sessions in London, which opened on 2 August 1898, in the basement of 31 Maiden Lane. The first London disk recorded was a rendering of *Comin' Thro' the Rye* by soprano Syria Lamonte.

To meet popular demand Berliner issued the first European disk catalogues on 16 November 1898. Listed in that pamphlet we find the first English piano recording, *Under the Double Eagle* made by a Mr Castle (8 August 1898); the first woodwind record, *Variations Brillantes* by the clarinettist of the Trocadero, A. A. Umbach (8 August 1898); and a first spoken-word recording, *Three Hard Questions*, a dialogue between Tom Birchmore and John Morton, made on 15 August 1898. No great names appear anywhere.

The first great name to record on disk was Clara Butt, the deep-voiced contralto of world fame. She sang in duet with baritone Kennerley Rumford at the Maiden Lane studios on 26 January 1899. Their piece was Goring Thomas's *Night Hymn At Sea*.

At this time none of the European disks were pressed in England. All the records emerged from the Hanover factory first set up in autumn 1898 and run by the Deutsche Grammophon AG, a name that persists until the present day. At Hanover the master-records were converted into metal stampers for mass production. Unfortunately, though, the acid etching system still used in fixing the traces on the masters, gave a roughness to the grooves. Eldridge Johnson found the ideal way around this defect. He recorded performances on thick wax disks, then sputtered the wax with an ultra-fine coating of gold and duplicated masters by electroplating. Further electroplating allowed him to grow many stampers from the masters and each record gained in quality from the smooth grooves provided by the original wax. This revolutionary change was first made on 14 May 1900; the first recording using wax was comedian George Graham's *The Colored Preacher*. When the new *Improved Gram-O-Phone Records* first appeared they were notable for their first use of the paper labels. Up until then all information had been scratched or printed on the shellac itself. But the record size remained unchanged at 7 inches.

On 3 January 1901 Johnson recorded his first 'Improved' 10-inch disk, *When Reuben Comes To Town*, sung by S. H. Dudley. It marked the end of the 7-inchers reign. At this time record buyers became aware for the first time, of the world's most famous trade mark, the dog and gramophone; the immortal *His Master's Voice*. When first painted, by Francis Barraud, the dog 'Nipper' was shown listening to an Edison cylinder machine. But the Gramophone Company offered to buy the picture if one of *their* machines was shown. So out went the phonograph and in came the brass-horned gramophone we all know. The altered picture was bought on 17 October 1899; it was first registered as a trade mark in the US in July 1900; first used in adverts in 1900 and first placed on record labels in 1902. The labels remained black and gold until colour came in with the first red labels in 1901. These were disks made by the great bass Chaliapin, known also to the irreverent as Charlie Appin.

As the quality of recordings improved so great performers overcame their doubts and came to face the recording horn. The first great composer to record was Edvard Grieg, who played some of his own piano pieces at the Paris recording studios of the Gramophone Co. at the end of April 1903. Despite the drawbacks of the system these first offerings are most moving to listen to.

Complete works whether orchestral, or vocal, were regarded as out of the question, but in 1903 the Gramophone Co. broke new ground by issuing the first recording of a complete opera. Its Italian branch issued forty single-sided 10-inch disks containing the whole of Verdi's opera *Ernani*. It also made the first disk recordings of the world's greatest tenor, Enrico Caruso, on 18 March 1902. One of these recordings was heard by Heinrich Conried of the Metropolitan Opera, New York, and Caruso was offered an engagement on the strength of that hearing. This is the first time that any artist had auditioned by sound recording. And it is the first success of such an audition. Caruso opened at The Met on 23 November 1903 and sang there for many years, thanks to that first disk.

Caruso's recordings went on to become top best-sellers; but the single record that first sold a million copies came from the world of the dance band. It was Paul Whiteman's *Whispering*, backed by *Japanese Sandman*, recorded for Victor in August 1920. It was made with his first band and was his first ever record.

One of the obvious drawbacks of all early disks was the short playing time. This was not too much of a problem for ballads or dance pieces, but all serious recordings were hamstrung. Even placing recordings on both sides of the platter was no real solution, it was still impractical to record most orchestral works as they were written. When Columbia issued its 12-inch record of Schubert's *Unfinished Symphony* it slashed chunks from the score in order to fit it on to two sides. It was their first symphony and when it appeared in July 1911 it seemed glorious, but it was a travesty when you consider that six 12-inch sides were really needed!

First of the brave efforts to increase playing times were made by the Neophone Company in 1904. It enlarged one brand of its disks to a giant 20-inches, and so lengthened the playing time to a maximum of ten minutes. At best this was a temporary solution. Not every record-player could handle such a large platter, and storage became even more trying. It was left to the 'Wizard of Menlo Park' to create the first viable solution to this vexing problem.

In 1927 Thomas Edison issued the first effective long-playing disks. These could run for forty minutes, a most remarkable feat considering that the disks revolved at 80rpm and were played acoustically. But the grooves on these disks were cut on the hill-and-dale system and the reproducing sound-box had to be driven by a lead-screw mechanism that was found only on the Edison Diamond Disk machines. It was a combination with a radically limited future. In the end the winning system was to be one based on a much slower speed and utilizing the lateral-cut grooves of the orthodox gramophone records.

The public heard its first long-playing 33.3 disks on 6 Aug 1926. These were the synchronized gramophone records used to provide the sound for the first programme of Vitaphone talking-pictures, shown in New York's 1,300-seat Warner Theatre. The strange speed had been arrived at in order to match each record with the running time of a 1,000-foot reel of film without sacrificing too much quality in the reproduction. Once this new speed had been fixed it was used as a standard by radio stations as well, and was the only speed considered for the first LPs when they were issued by Victor in 1931. Their first recording, demonstrated on 7 September 1931, was of Beethoven's *Fifth Symphony*.

It was a premature move; the technical problems were too great and the reproducing machines were too expensive, the cheapest was £100. The venture collapsed within a year and it took seventeen years before the LP was remarketed by Victor's rival, Columbia. The Columbia LPs of 21 June 1948, differed from the earlier records. For the first time they offered the slow speed coupled with microgrooves. It was a winning combination enhanced by the use of vinylite as a silent surface. But before LPs became acceptable the problem of quality had to be tackled.

From the birth of sound recording the results were always limited by the lack of amplifiers and microphones. All the sound in the early studios had to be captured by large horns and fed to cutting heads that were purely mechanical. They were more like the old sound boxes seen on early gramophones than anything else. They could not harness all the sound power in the studio and the frequency range was inadequate. And then there were sundry distortions to cope with. Only a sensitive system using electrical techniques could make the advances to high fidelity. Electrical methods had been thought of from the very first. As early as 1886 Bell and Tainter's Volta Labs had patented an electrical recording head meant to engrave a track on wax cylinders. And in 1911 E. H. Amet first patented an electrical method of recording multitracks. But all the early concepts were shelved because of the lack of amplification.

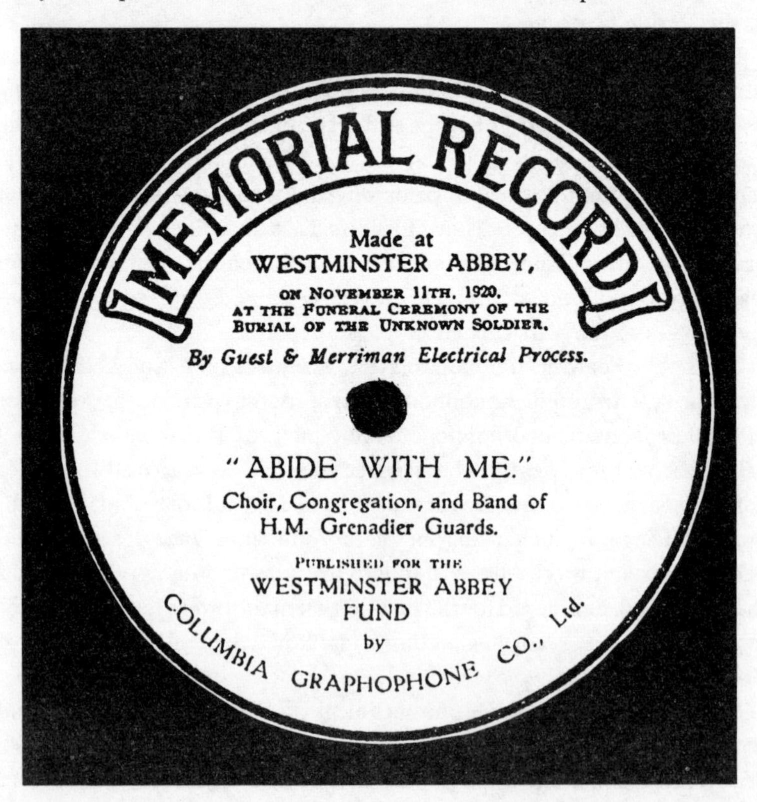

The label of the first commercial electric recording

A claim to have been first, to have solved the amplifier problem has been made by the late Arthur Kingston, a veteran of talking pictures and a distinguished inventor. He has stated that in France, '... We started experimenting. We made microphonic relay amplifiers – one microphone feeding another, feeding another – and the results we got were so good that we went straight ... to direct electric recording. So we were doing electric recordings in France on talking pictures in 1909. The quality of sound was so good that my boss got a contract from the French Columbia company to make disks of the Garde Républicaine.' Unfortunately no trace has yet been found of any of the French band records mentioned by Kingston, so the first commercially-issued electric record has to be rated as an English Columbia disk of 1920. This really was an oddity. It was part of the burial service of the 'Unknown Warrior' recorded in Westminster Abbey by H. O. Merriman and Lionel Guest. The sound was of appalling quality, and the disk had novelty or sentimental value only, but today it is a sought after collectors' piece and the copper matrix is treasured by the British Museum. At the start of 1923 the Merriman-Guest recording system was first used by Columbia Records (US) to make test records. But the US firm went bankrupt and the electrical revolution was quashed.

Meanwhile, Western Electric's Bell Labs married public-address amplifiers to a condenser microphone invented by E. C. Wente and created the first masterly electrical recordings. The microphone had a wide, smooth response and the amplifier gave enough boost to operate the electromagnetic recording head. This well-matched combination of firsts gave rise to the full-blooded revolution in recorded sound that first made its debut in the cinema.

By 1925 the Western Electric system was in the hands of both Victor (US) and Columbia. Victor issued its first electrical disk in April 1925. It featured the mid-March location recording of the University of Philadelphia's Mask and Wig Club, coupled with a rendering of *Let it Rain, Let it Pour* played by Le Paradis Band. Columbia's first electric disk made more impact with its recording of the unison singing of 4,850 voices of the fifteen glee clubs assembled in the New York Metropolitan Opera House, on 31 March 1925. These were the two great firsts that people marvelled at for their clarity and sheer power. But for all that, these giants were beaten into second places by a pigmy outfit in Chicago. This obscure Marsh Recording Labs had issued the first electrically generated commercial entertainment recordings as early as the autumn of 1924. These *Autograph* records were of jazz performances and featured Jelly Roll Morton, King Oliver and other jazz stars. Yet they were not sold widely and were overlooked by most historians.

Electrical recordings were ignominiously fed into a market quite unable to appreciate their full glory. Most people owned acoustic record players using mechanical sound boxes and horns. The horns, whether external or internal, cramped the frequency range and by, adding unnatural peaks, coloured the sound. The sound boxes cramped and coloured as well. Using such equipment could only give a hint at the real beauty and power captured in the grooves. These machines were incapable of handling the first electrically recorded symphony, first released in December 1925. This was Tchaikovsky's No. 4, played by the Royal Albert Hall Orchestra under Landon Ronald.

To match the electric records one needed an electric reproducing unit and the first such units of 1926 used electric pick-ups meant to be plugged in to the amplifiers of radio sets. Then Brunswick issued its *Panatrope*, the first all-electric reproducing machine. It was the supreme audio marvel of 1926, though the claims made for it were exaggerated. It was said to be able to reproduce from 16 to 21,000 Hz, but this was sheer moonshine. The heavy moving-iron pickup was incapable of such flexibility. Contemporary tests showed the machine to be thin at the top and heavily boosted at the bass end. Despite this it was a great step forward.

Brunswick made an even greater advance when it introduced the first light-ray recording system in 1926. It used a crystal mirror weighing a trifling 5mg, that picked up the sound vibrations and reflected a fluctuating light beam on to a photo-electric cell. The cell converted the light rays into current and fed this to an amplifier, which in turn passed its output to a moving-iron recorder. This was the first claim to have reached the goal of high fidelity. It was this device that handled the range of frequencies from 16 to 21,000 HZ, for the first time. The electric player was soon joined by the first marketed auto-changer of 1927. This was a Victor product based on experiments dating back to 1920. Other manufacturers waded in with their own clever changers, and some proved just too clever for the health of the shuffled records! But most people stuck with the safer and much cheaper method of hand changing.

The electrical revolution meant drastic changes in recording studios. The old, cramped and dingy rooms were inadequate for the new techniques. Now it was possible to have a complete symphony orchestra in front of the recording microphone, instead of the old system, where an orchestra was a score or more players huddled together around the dominant sound-collecting horn. So in 1931 the first studios meant specifically for sound recording in Britain were built. On 12 November Sir Edward Elgar formally opened the famous Abbey Road studios of HMV and a brilliant new era began.

By this time, radio had become a competitor in the field of home entertainment. At the same time it became an ally of sorts. A record played on air was sure to catch on. And the first disk-jockeys began to emerge. (The very first DJ is counted as Christopher Stone, who began introducing the latest disks back in July 1927, from the BBC studios at Savoy Hill, London.)

Record speeds remained static, though, at 78rpm. The LPs never caught on and a slower speed of 24rpm (first used in 1924) was intended for speech only and restricted to Talking Books for the Blind. But researchers were still looking far ahead and in 1931 A. D. Blumlein of EMI took out his first patent for stereophonic recording and disk-cutting techniques. These techniques were first applied commercially years later, in 1957. Earlier attempts at stereo disks count mainly as amusing novelties, they had no long-term future ahead of them. This is obvious when you look back at the very first stereo disks issued by Pathé about 1912. They used twin tracks, side by side and twin pick-up arms to traverse the grooves. This meant buying a large and expensive machine simply to hear the limited range of recordings on offer. But viable stereo disks had to wait for the full development of the long player and the introduction of vinylite. Vinylite was first adopted as an emergency material during the Second World War. Only the shortage of the customary shellac led to the search for a substitute and velvety-smooth vinyl was found to be outstanding, though only a light-weight pick-up could do it justice.

When Columbia's first vinylite long-players were announced, on 26 June 1948, they were standardized at the 33.3rpm speed used on the short-lived RCA LPs of 1931. But there was a crucial difference. For the first time the old coarse, lateral-cut grooves were replaced by microgrooves, giving from 224 to 300 grooves per inch. RCA though invited to join the launch, stayed aloof and in January 1949 it introduced a rival system, the first 45rpm disks. These went back to the old-time size of 7-inches, but by using microgrooves, gave the same playing time as a 12-inch 78rpm record. At the same launch RCA unveiled the first machine designed to handle their 45s. It was a neat autochanger with a 1.5-inch centre spindle, so nothing but one of their disks could ever be used on it. And the large hole in their disks meant that no other machine could take them. It was a system doomed to take a poor second place to the Columbia venture.

Long players gave a death blow to the 78 standard and then stereo recording began to oust the mono microgrooves. Demonstration stereo disks were first played at the Los Angeles trade reception in September 1957. Then, in May 1958 the first stereo disks designed for single pick-ups were marketed in the USA. By June, the first stereo disks from Pye were on sale in the UK. By 1960 the first fake stereo pressings were on offer. These were old mono records re-channelled to compete with the new sound experience. By 1969 the whole of the record industry had switched over to stereo exclusively. But the attempt to go further and introduce four channel or quadrophonic records was a dismal failure. The first quad recordings were those issued by Acoustic Research of Boston in 1968. The larger CBS demonstrated its own system in 1971, then other systems entered the race. But it was a three-legged race to extinction. Four speakers, special amplifiers and a limited repertoire never had much appeal.

It was different with the introduction of digital recording. This system was fully compatible with orthodox stereo players and pick-ups, but it offered a cleaner, more realistic sound. The first digital recording of note was Decca's 'New Year's Day Concert in Vienna', a live recording of the Vienna Philharmonic, made on 1 January 1979 and released in April. It was honoured by being the first recording to gain the US Electronic and Engineering Award. Yet if the sound was kept undergraded at source, it still had to be delivered to the listener through a stylus running in a groove and exerting pressure. The compact disk was then developed to free the record from all the drawbacks of wear and tear.

The compact disk was developed jointly by Philips of Holland and Sony of Japan. Its arrival was announced on 17 May 1978; but the CD hardware and records were not launched until October 1982 in Japan, and March 1983 in Europe. The disks were shown to double the playing time of the best LPs. And they were smaller; just 120mm in diameter; and so light. But, more remarkably, they were virtually indestructible. The recorded information was carried in microscopic pits coated with aluminium. Then this information was sealed under a transparent plastic layer which gave it complete protection while allowing the pits to be read by a tracking laser beam. With all these incredible advances it was certain that the CD was the record destined to supplant all others.

And this is exactly how things are shaping up. The initial high cost of CD players and disks is now behind us. Everything is affordable and high-tech has triumphed to the point where the same CD player can even display pictures through a television set or monitor, the pictures themselves being carried on a disk that looks exactly like a standard CD. It looks as if the system is here to stay. Unless . . .

MAGNETIC RECORDING

The first *patent* describing a system for recording sound magnetically was issued to Dutchman Wilhelm Hedick in 1888 (No. 569). It was a complex method, way in advance of the technology of the time. An easier method was outlined in the same year by American Oberlin Smith. He hit on the idea of recording on cloth bands interwoven with metal threads; but neither of these ideas went beyond the paper stage. The world's first practical magnetic recorder was created in 1887 by Valdemar Poulsen the inventor who came to be dubbed 'The Danish Edison'. Patented on 1 December 1898, it was known as the *Telegraphone*, and the first models recorded on piano-wire. Later on other styles using steel bands and even

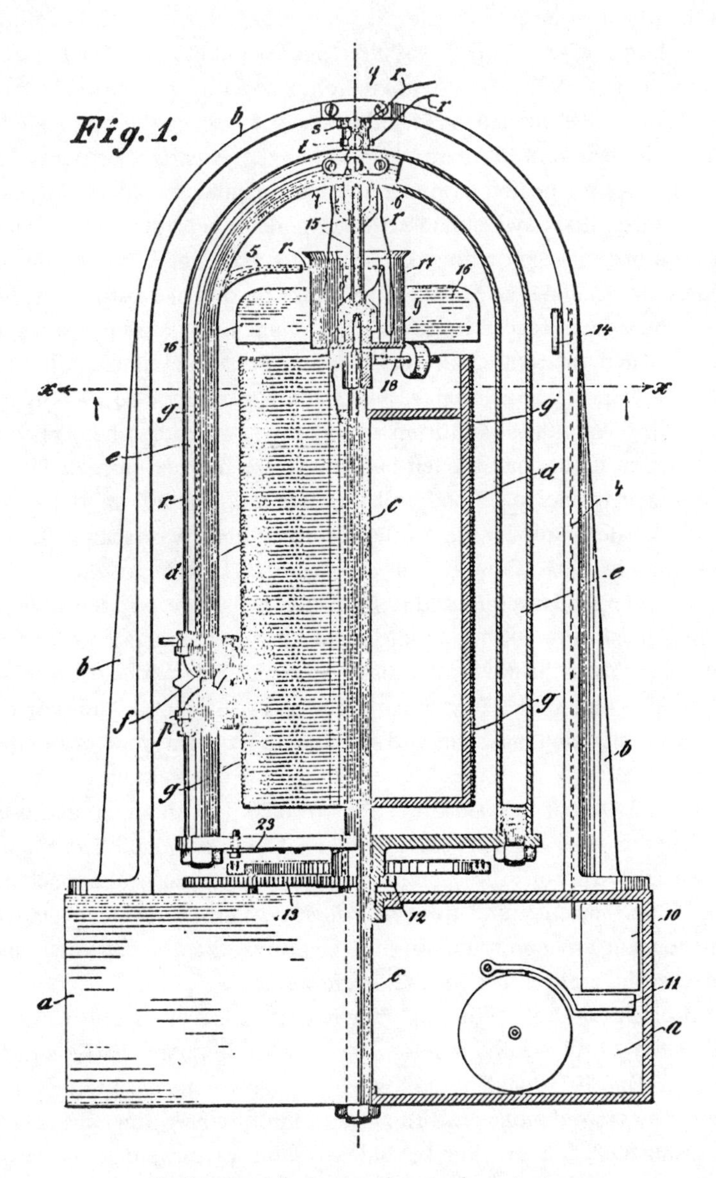

Poulsen's original magnetic-recorder patent

steel disks were made but the wire-recorder was the pattern that showed true commercial promise. Unfortunately at a crucial stage of its development it lacked a means of amplifying the sound and this stifled its progress but its basic principles were those which came to govern all future developments in tape recording including video. And Poulsen can be credited with yet another first, since his Telegraphone was originally intended to record telephone conversations or messages. It qualifies then as the first answering machine.

Poulsen's wire-recorder was first put to public use in Paris at the Exposition of 1900. Many visitors were permitted to record their comments and have the delight (or otherwise) of hearing their pieces played back within minutes. At this Exposition the Austrian Emperor Franz Joseph became the first monarch to use a magnetic recorder. The recording he made that day is still in existence and has been dubbed and used in many a broadcast programme. The quality is dreadful, but within just a year, new machines gave results comparable to that of a distant telephone. Now, they could record for at least 30 minutes, unlike the 30 seconds of the 1900 demonstration machine. This new Poulsen model kept the name Telegraphone but was in all respects a new venture. The 1900 machine used wire wound round a drum and a moving recording head. The new production model had a fixed head and wire fed reel-to-reel. This was the first machine to embody all the essential features of the modern tape-recorder. Only biasing of the recording head was lacking. This was remedied in 1902 when the first patent for DC biasing of the recording head was applied for, (granted 1907). The biasing improved sensitivity and reduced distortion of the sound. Another Poulsen patent is worth noting since it involved the use of steel disks with recorded tracks *adjacent to each other*. This patent of 1903 marks the first use of a practice later borrowed for office dictating machines and still used in recording data on computer disks.

In 1912 Doctor Lee De Forest borrowed a Telegraphone and used it to develop several electronic amplifiers. These first amplified machines are actually shown in his patent of 1913. He even used the machine to add synchronized sound to some *Biograph* films before changing tack and concentrating on sound-on-film. De Forest then, was the first man to forecast the great future for the invention.

After a period of stagnation, new important developments came with the issue of two US patents. The first to Leonard F. Fuller, (No. 1,459,202) in 1918, dealt with the use of a high-frequency alternating current to erase redundant recordings. Up until then erasing was achieved by using DC, but this left behind a residue of unwanted background noise; the new AC system gave a clean result. This was the first use of a technique that was essential for good quality sound. Just as important for sound quality was the 1921 patent (No. 1,640,881) granted to W. L. Carlson and Glen Carpenter. They overcame the limitations of the DC bias to the recording head by using an AC method. This again was the first use of a method that became essential in later years.

But despite all these fresh developments, magnetic recorders still stayed wedded to the office. The first man to take them out of these restricted bonds was German Dr Kurt Stille. In 1928 he developed a large machine recording on steel tape and easily synchronized with the sprocket drive of a film projector (No. 319,681). He sold his patent rights to Ludwig Blattner who was intent on using the machine to make sound movies. Blattner owned the 'Blattner Colour and Sound Studios' at Elstree, and there he produced the first practical Stille machine and promptly renamed it *The Blattnerphone*. This first machine was demonstrated to the press on 10 October 1929. The first item played back was a monologue by Henry Ainley. Then the first magnetic sound talkie was shown; it was an unambitious short in which Miss Ivy St Helier sang at the piano and then gave a witty talk, but it impressed the audience. The machine certainly had a future. And the BBC saw this very soon, for they swiftly bought the rights to an improved form of the machine and began to use it to record broadcasts for retransmission.

At first these were unheralded affairs; publicity arose only when it was first used by Lord Reith on the first day of the Empire Service in 1932. The steel-tape recording meant that the ceremony could be sent out at different times of the day and night to accommodate the time differences in the far-flung Empire. At this time we have the first production of gramophone records from tape-recorded masters. This came about after an accidental erasure of part of the 1932 Christmas Day feature programme. As a safeguard, the BBC then contracted with the British Homophone Company to record selections on to disk from their Blattnerphone tapes. The excerpts were fed by land-line from the Maida Vale Studios to the record plant at Kilburn.

In 1933 the Marconi Company bought all rights in the Blattnerphone and they then created a new version known as the *Marconi-Stille* machine. These first went into service with the BBC in 1934 and coped well for over thirteen years. In that time, they were the source of many legends and some tall stories but it *is* true that the steel tapes ran at such a fast speed that engineers cowered in far corners if they feared a tape breakage. And it is true that the edited joins in the tapes were hand-soldered. This meant that the Marconi machines were the first to need five heads. Two were there to be instantly switched into position every time a soldered splice passed; for in passing, the join destroyed the knife-edged pole pieces of the record and replay heads. A whole tray of replacements had to be ready to cope with the heavy toll on pole pieces.

It is of course pure fiction that the first murder at Broadcasting House was also the first solved with the aid of a Blattnerphone. The fiction was created by BBC producer Val Gielgud and Holt Marvell in their book *Death at Broadcasting House*. They depicted the murder of a radio actor at the very time when he had just finished his small speaking part. In struggling for clues the Yard detective learns that the play had been Blattnerphoned for use on the Empire service. '"We can hear that scene" said producer Caird, "not only over again, but over and over again. As often as you like. I wonder if the murderer thought of that?"' He hadn't and the machine trapped him! This detective yarn, first published in 1934, became the basis of a film in which there is much talk about the Blattnerphone but not a single shot of it. Still the film *is* the first to star a tape recorder, even if it is a shy, retiring one.

Yet another German development brought about the demise of the steel-tape recorder. In 1920, A. Nasavischwily first suggested using a powdered magnetic material, instead of solid metal, as the recording surface. Then in 1927 Doctor Fritz Pfleumer first began to search for a flexible recording tape made up of an insulated base and a magnetic coating. It was a long haul and he only filed for a patent in February 1929. But his first patent showed that he was on the right track. In 1930 he joined forces with AEG of Berlin, who then brought in the chemical expertise of I. G. Farbenindustrie of Ludwigshaven. AEG itself, concentrated on a machine to handle the new tape, which was the first using a cellulose acetate base, coated with a carbonyl iron powder. The first batch of this tape was produced in July 1934. But the first demonstration of the AEG machine in August 1934 disclosed unexpected problems. It took a mere eight weeks to overcome these problems and in October 1934 the revised model had its first day of triumph. More revisions led to the market model first shown at the Berlin Radio Show of 1935, where its excellence was recognized at once and all the available recorders were sold on the spot. The tape recorder as we know it had arrived. Even so, not everyone found the recording standards high enough. When Sir Thomas Beecham recorded the first public concert on these new machines (in Ludwigshaven, 19 November 1936), he found the quality far below the standard he expected from a gramophone disk. On listening to the tape of that first concert one has to agree. The woodwind tone in particular is thickened and the dynamics are compressed. AEG were anxious to reach higher levels, since they wanted to interest broadcasting companies, and they worked on their *Magnetophones* (as they named them) until they made them fine enough for the Reichs-Rundfunk Gesellschaft to adopt them for radio use. Their first such use began in January 1938. By then, the tape had been redesigned with a brown gamma-ferric oxide coating, the first to permit high-frequency recordings.

When war came, a blanket of secrecy fell over fresh German developments. But in the USA the Brush Development Company began marketing the first black-oxide-coated paper recording tapes in 1939. Yet the emphasis was still on recording on wire or steel tapes, as demonstrated at the 1939 World's Fair, where Bell Telephones displayed the first tape machine able to record and reproduce stereo sound. The effect was impressive, but it was dependent on twin Vicalloy steel tapes costing $1.50 a foot.

While the USA was still busily developing wire recorders for the armed forces, the Germans were making portable tape machines for their forces. And Magnetophones were utilized in all the main radio stations. The sound quality of these broadcasting machines became so good that it grew difficult and often impossible to know if the broadcast was live or recorded.

In July 1945 an American soldier stopped his jeep at a crossroads. To the right lay the road to his HQ in Paris; to the left was the road to Frankfurt. He thought hard, turned his back on the bright lights of Paris and headed off to Frankfurt. The first step in creating the American tape recording revolution had been taken. For John Mullin it proved to be the greatest decision of his life. He had chosen to track down a legendary machine that he had heard of. It was said to be kept in the castle at Bad Nauheim, a place once used as a temporary radio station by the Germans. When he reached the castle he found the machine gleaming and intact. Though it looked like other Magnetophones, it had an astonishing frequency range, unlike any other machine Mullin had ever heard. It was the first recorder to match live sound and no disk recorder on Earth could compete with its realism.

Mullin's quest began there and then. He tracked down three more of these superb machines; presented two of them to the Signal Corps and commandeered two for himself. But how to get the brutes back to the States? He invoked the 'War Souvenir' regulations. If such a souvenir could fit into a mailbag in Paris, then it qualified. He photographed everything; made schematic drawings; took the machines to pieces and fitted them into thirty-five wooden boxes. Each box filled a mailbag and the whole lot was shipped over to San Francisco.

Within four months the machines were rewired and reassembled. In May 1946 Mullin gave the first public demonstration in the USA of this King of tape recorders. At that stage only professionals took an interest, the domestic stage had not yet arrived. But one professional showed more than average interest, and his intervention was electric. Bing Crosby was looking for a way to record his radio programmes, and after changing to ABC he got his way and began using 16-inch transcription disks. But when these disks were edited the loss in quality showed, they became fuzzy. However tape could be edited simply by cutting and splicing. The original high quality remained. And that is what Crosby wanted so he called in Mullin and his machines, and in August 1947 made first recording on tape: the song *Blue of the Night.*

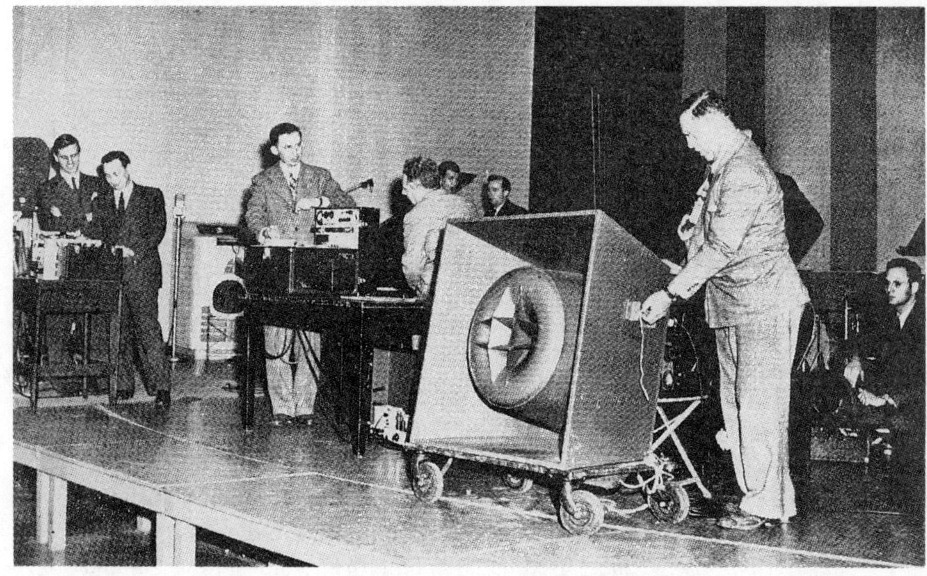

The first public demonstration of Mullin's revolutionary tape recorder in the USA

He followed this with patter and another song; loved what he heard played back; then dropped the transcription disks and became the first star of note to rely on tape recordings. As well as that, he backed the magnetic system with his millions and a new industry was born. It made a giant out of Ampex (the first firm to improve on the Magnetophone) and it sent the 3M company (who made the tapes) to dizzy heights.

The first of the new Ampex recorders, Model 200, was delivered in April 1948. The first professional order, for twelve machines, was placed by Bing Crosby Enterprises; and the first Ampex advert carried the heading 'Here's the machine that put Bing Crosby on tape . . .' A little later in 1948, Mullin gave the first successful demonstration of the art of using tapes to create top quality gramophone records. He fed two recordings of Bing Crosby's songs to the record-cutting lathes at Decca Records. The results were so convincing that Decca and Capitol both installed Ampex machines for creating their masters, and the rest of the record industry began to rethink its techniques. At the same time came developments in the domestic market when Magnecord switched from making wire recorders and made a portable tape recorder instead. In 1949 Magnecord went one further step ahead when it demonstrated the first stereo recorder; both tracks were recorded on one tape, unlike earlier attempts. Many of the strides now taken by the industry were made possible by a new red oxide tape with superior recording characteristics developed by Minnesota Mining. This tape, No. 111, was first demonstrated in 1947, and first marketed in 1948. By 1950 there were over twenty-five different tape recorders on sale in the USA. There was even a network of tape exchangers, called Tape Respondents International, and a body selling pre-recorded tapes. This was Recording Associates (US) whose first catalogue listed a grand total of eight tapes.

By 1951 Crosby Enterprises was able to demonstrate the first tape machine able to record video signals. One year later the Victor Animatograph Corporation brought out its *Magnesound* device which converted movie projectors using optical tracks, over to machines able to use film striped with a magnetic oxide track. And the first taped language teaching courses were placed on sale by the Tape Correspondence School. Its first course was in French.

The boom continued in 1953, with the first issue of the first magazine completely devoted to the new art, called *Magnetic Film and Tape Recording*. Then the manufacturers formed their first special body, 'The Magnetic Recording Industry Association'. High-speed duplication was first introduced by Terry Moss and RCA announced their first video-tape recordings in colour.

By 1954 tape recorder models proliferated. Minnesota Mining catered for the new users by offering the first Extra-Play tapes (No. 190) on a 1mil acetate base. They also introduced their first tapes on a polyester base. Rivals then countered with the first *Mylar* based tapes, first as standard play, quickly followed by an extra-play tape of 1mil from Reeves Soundcraft. 1954 was further marked by the gaining of an Oscar for the magnetic sound in the Cinemascope production *The Robe*. This went to Reeves Soundcraft.

The search for even longer playing tapes hotted up in 1955. First in the lead was Orradio Industries, who issued the first Double-Play tapes on a half-mil Mylar base. Machines, ranging from the first tiny pocket tape recorder, by Broadcast Equipment Specialities, to office dictating machines, like the *Stenorette* and to huge multi-track consoles, now sold easily. And giant concerns, like Bell and Howell, and Columbia Records, were alerted and made their first entries into recorder manufacturing. Even Westinghouse used the new technology to develop the *Talking Elevator* which first went into service in 1956. By 1957 the first recorded book on tape was on sale. Fittingly enough it was called, *All About Tape on Tape*, and was published by Tape Recording Magazine. It was, like all the other pre-recorded tapes of the time, meant for a reel-to-reel machine, and this always meant a great deal of fuss and bother for many people. Cleveland inventor, George Eash looked for a simpler way of using tape and made the first cassette for holding tapes ready for replay. He thought first and foremost, of the car user. Handling reels in a car could prove frustrating, even dangerous, but a self-contained, ready-threaded tape unit would reduce things to child's play. His first prototype cassette and DC machine to handle it was ready in autumn 1956. By 1960 it was in use and large record companies were co-operating in supplying tape licences. But it was far from the long-term answer to in-car entertainment. That was not adequately met until 1965, when Bill Lear persuaded the Ford Motor Company to offer cars with built-in cassette players able to handle the new Lear Jet cartridge. A Jet cartridge used eight tracks (its rival Muntz used four) and had an automatic switching method.

All the new thoughts about cassettes were overshadowed by the presence of the Philips mini-cassette first developed in 1961 and first unveiled in Berlin in 1963; along with the first player able to use them. This is the ubiquitous audio-cassette we all know so well. Its widespread adoption is due to a shrewd move by Philips, who gave their patents free of charge to any manufacturer who wanted to join in producing the mini-cassette. The micro-cassette, first introduced by the Japanese company Olympus in 1976, has as yet offered no threat to the Philips pattern. Its smaller size, and greatly reduced speed makes it ideal for pocket-size audio-pads, but limits its use for any type of music. The real future seems to lie with the increasing use of digital recording and playback techniques. The first audio tapes (DATs) along these lines were marketed by Aiwa of Japan in February 1987. Two competing machine systems emerged at this time. JVC's first S-DAT used a fixed head, while Sony and Matsushita used rotating heads. The rotating head idea is the one that beguiles most manufacturers at present. It gives a result, they claim, equal to that of the compact disk. But in the end, it is price and versatility that will decide the issue, not manufacturers' hyperbole.

Alongside all these easily noticeable advances went the less publicized developments in computers. The first computer, being conceived in 1822, arrived well before the days of magnetic records. This was the Analytical Engine partially constructed by Charles Babbage, who worked on it for over forty-nine years. It was way ahead of its time, but, had it been completed up to the high standards of precision needed, it could have reached the standards attained by the early electronic computers. Since it has been celebrated on a British postage stamp, it becomes the first computer to be pictured in the history of philately.

The electrical computer era began with Doctor Hermann Hollerith's machine of January 1889. Constructed in New York, it used punched cards to feed information to electrical circuits which then tabulated the results at high speed. It was first used to analyse the results of the 1890

The first use of the Hollerith Tabulator in the US Census of 1890

US Census. Its punch card method became an invaluable asset to calculator designers from then on. The first all-electronic computer, though, was not seen until 1946. This was the US Army Ordnance ENIAC machine. It used 18,000 vacuum tubes and semi-conductor diodes; developed a great deal of heat but efficiently dealt with its job of calculating firing tables for the artillery. This machine, developed at the Ballistic Research Labs, in Aberdeen, Maryland owed a good deal to earlier work carried out by Professor John Vincent Atanasoff of Iowa State College. Atanasoff's ABC machine, first made in 1939, differed from ENIAC by using a binary system (ENIAC was decimal). ENIAC's great advantage lay in its ability to be programmed, but it managed without tape. The first of the new breed of computers to use tape was the Harvard Mark 1 of 1944, a product of IBM. It was the first machine to use the register system which stored bits of data which could represent instructions, letters of the alphabet, or binary numbers. But it was not until 1948 that computer programmes were first stored in the computer memory. This was a revolutionary departure led by the EDVAC machine built on ideas first formulated by John von Neumann. The first of these machines was built at Manchester University. As the UNIVAC 1 it was first marketed in 1950. From then on the developments were speedy and many. Transistors, first developed by William Shockley in 1948, came to replace the space-consuming valves; magnetic tapes replaced the punched paper tapes and punch cards; until they, in turn, were replaced by magnetic disks. Today we have reached the stage where magnetic software makes it possible for a desktop computer to surpass the giant mainframe computers of the 50s and 60s. They are now faster, with more power and memory and much more adaptable. Tomorrow's lap-top portables are sure to reach these standards and more. Only the printers show no signs of shrinking in size – as yet.

• TOOLS OF CONFLICT •

Explosive force has dominated armed conflict for centuries. For good or evil it is now the prime factor that determines global security or destruction. Compared with Man's history though, it is relatively new on the scene. Yes, the Chinese used war rockets in the thirteenth century, but these were marginal weapons, of no real importance. Only with the birth of the cannon can we say that the era of explosives began. The first reliable record of a cannon is found in an English manuscript of 1326. On one of the illuminated pages of *De Officiis Regnum* (*About the Duties of Kings*) is shown a vase-shaped cannon being fired with a hot iron. Its projectile is shown to be a large arrow. An arrow was certainly a sensible choice since the first cannon were hardly strong enough to take a close fitting projectile. Despite this we do have a fine illustration of a gunner firing a cannon ball as early as 1400. This first recorded use of round-shot is shown in *Belli Fortis*, a German manuscript by Conrad Leiser. He also shows the first-known cannon to have a straight-sided barrel.

Which type of cannon was used at the Battle of Crécy (1346) is not recorded, but at least five pieces roared out that day and 'struck terror into the French Army'. This is the first recorded English use. But the French had first employed a cannon aboard a ship of its Harfleur fleet in 1338. It was an iron barrelled model that fired large bolts, similar to the crossbow bolts. Reduced size cannons were the basis of the first hand guns as we know from the earliest known specimen, the Tannenberg gun found in Tannenberg Castle, Hesse. Cast in bronze, it is about 12.5-inches long with a bore of 1.75-inches. Like all those early guns it would have been plagued by the shortcomings of the early gunpowder mixtures. Gunpowder in its first forms tended to settle into a hard mass and this made it erratic in use, since an igniting flame could not work through the mass speedily and predictably. The first steps to remedy this were taken in France. The standard mixtures were soaked in water, dried into cakes and then rubbed through a sieve to give a consistent size of grain. This evenly grained powder allowed the flame to pass freely through each charge. As a result the firing was virtually instantaneous, the force was increased and the results became predictable. A given weight of dry powder would always give the same power, in the same gun.

From those early beginnings and for centuries to come, all guns, whether field-pieces, ship-borne, or hand-held were fired by sparks, flames or hot irons. And there were many clever mechanisms devised just to bring fire to powder; starting with the first matchlocks of 1411; progressing to the first wheel-locks of 1530; and culminating in the first flintlocks of 1547. All this inventiveness centred around the problems set by the use of loose powder.

The solution was slow in coming. The first step towards the modern cartridge method only came after the introduction of fulminating substances. These are compounds and mixtures which explode when struck. They were first discovered by Edward Howard in 1799, but because they had to be handled with great care they were not put to military use at once. Their first use came on the gaming field in 1807. Their first user was the Reverend Alexander Forsyth, the trigger-happy vicar of Belhevie, in Scotland. Sparking flints were vulnerable in damp weather, so he adapted his flintlock to take a revolving magazine filled with his fulminating mixture. A twist on the mechanism deposited a small amount of the mixture in the tube leading into the gun barrel. A pull on the trigger drove a rod down hard on the mixture; this fired, and ignited the charge in the gun. Our vicar's example inspired the invention of the first percussion cap in 1814. Now the fulminating mixture was packed hard (when wet) into a small copper cap. The cap was placed on the end of a hollow metal nipple that led into the gun barrel. The fall of the gun's hammer crushed the cap down on to the nipple's end, ignited the mixture and fired the gun.

The logical next step was to fix a percussion cap into a self-contained cartridge and develop a gun able to handle it. This is exactly what the Prussian gunsmith, Johann von Dreyse did. His gun was made as a breech-loading model; his cartridge had the percussion cap fitted well inside the casing; and the trigger released a spring-powered needle which pierced the casing and hit the cap. This brilliant design of his (first shown 1841) was the first breech-loader adopted by the Prussian army in 1848. It was also the first gun to use the bolt-action, later adopted by every manufacturer. When news of the needle-gun reached the French they began to search for a gun to match it. And by 1863 they were able to demonstrate their first Chassepot rifle. This used a much shorter firing-pin and a cartridge with the cap very near the base. This was a step nearer to the self-contained all-metal cartridge that came to give birth to the machine-gun, the automatic pistol, and all the major rifle designs in the world. The first all-metal cartridges were displayed in 1865. In England Boxer and Daw showed their first pattern and in France Gosselin and Schneider (independently) showed their patterns. Both Boxer and Schneider were first to show the value of centre priming as opposed to the awkward rim-fire design. When centre-fire became coupled with one-piece brass cartridge cases (1867), the repeating rifle became possible. But the repeating *pistol* or revolver had been first made to work without metal cartridges. The idea itself is an ancient one. There is a horse pistol in the Tower of London which has a rotating device, and that dates from 1680. And in 1814 John Thomson of London devised a flint-lock pistol with a nine-shot revolving magazine. But the revolver only became popular when the metal cartridge made it easy to use and reload. No one man made that possible but the first to popularize the revolver was Samuel Colt whose first patent was registered on 22 October 1835.

At first all revolvers had to be hand-cocked and this clumsy feature led Robert Adams to make the first self-cocking revolver in 1851. Colonel G. V. Fosberry of the British Army went one step further with his self-setting and self-cocking pistol. On 16 August 1895 he patented the first automatic revolver. It used the force of the recoil to move the cylinder to the next chamber and cock the hammer at the same time. The automatic pistol utilized the recoil as well, but it dispensed with the cylinder and used a flat magazine for the bullets. The first automatic was the work of Austrian Joseph Laumann in 1892. His pistol was automatically loaded and cocked but turned out to be unreliable after a little use. The first reliable automatic became the fine model made by Hugo Borchardt in 1890. This became the pattern that was drawn on to create the first of the famous Luger pistols of 1898.

Harnessing the force of recoil became an essential feature of machine-gun design, but even here the first of the breed managed without it. They used a mechanical method of reloading and firing. First of these was Wilson Agar's so-called 'Coffee-Mill' gun. The turn of a crank brought bullets from an overhead hopper into the firing chamber; tripped the firing hammer; threw out the spent cartridge tube, and reloaded the next round. Agar's gun of 1862 was first used by the Union Army in the US Civil War. The same war produced the most famous and most reliable of all the mechanical machine-guns, the *Gatling Gun*. In 1862 Richard Jordan Gatling made his first model by fixing six rifle barrels in a revolving frame. As the frame rotated it brought the firing and loading mechanisms of each barrel into contact with a series of fixed cams. One cam opened the bolt fitted to each barrel, the bullet fell into its chamber, and another cam closed the bolt; yet another cam released the firing pin and a final cam opened the bolt, throwing out the spent case. The cycle then recommenced. It was a much-valued weapon until superseded by the faster automatic machine-guns. Even so its principle is *still* in use today on quick-firing anti-missile guns used on naval vessels, and in the Gatling cannon fitted to the A-10 Avenger tank.

Automatic machine-guns came to use all the wasted energy found in guns. As well as the recoil power, they used the gas pressure from the propellant, and the first to capture some of this wayward energy was the American inventor Hiram Maxim. Though an American his military research was conducted in Britain, where in 1885, he demonstrated the first automatic machine-gun to British Army observers. It used a belt-feed for the bullets, recoil power to activate it, and could deliver an amazing 600-rounds-per minute. It was soon adopted in Britain, Russia, Germany and Italy. Maxim's success caused others to look at machine-guns and in 1890 John Moses Browning (USA) perfected an automatic system that used the propelling gas as the source of power. This first gas-operated machine-gun was taken on board by the Colt Company, and, after modification by them, was the first machine-gun adopted by the US Navy (1895).

Though the belt-feed was unsuited to artillery some aspects of machine-gun technology did appeal to those searching for a way to make field-guns quick-firing. The problem there lay in the light weight of these guns. A garrison gun, or naval gun, could be bedded down to a great weight to absorb the shock of recoil. The field gun, by contrast, would leap backwards after each shot, preventing fast reloading and aiming. The problem was solved by the French artillery experts at their Puteaux arsenal when, in 1897 they produced the first quick-firing field-gun. They used a quick-acting breech mechanism; but of much more importance, they used the recoil power to drive pistons that were part of a complex hydro-pneumatic braking system. This allowed the gun to stay still after each round; reloading was swift. This 'French 75' was first shown at the manoeuvres of 1900, but not fired. Its secrets were closely guarded for years, but eventually its sound principles became adopted for quick-firing guns everywhere.

The larger artillery pieces then came to adopt the hydro-pneumatic technology, but for them the greatest advance had come much earlier with the advent of breech-loading. Muzzle-loading seemed fine in the days when powders were fast burning. But in the 1870s slow-burning gunpowder came in. It delivered more power to a shell but needed longer gun barrels to develop that power.

Longer barrels were harder to muzzle-load, but breech-loading would make the task fast and easy. It also meant that the new advances in small arms ammunition could be used for even huge guns. Early breech-loading ideas had been put forward in 1821 (Lieutenant Croly), and in 1846; but the first breech-loader to be produced was the 1,000-pounder Krupps gun, first publicly shown at the Paris Exhibition of 1867. Ironically it became the first gun to bombard Paris three years later! To improve performance, the gun barrel was given internal spiral grooves.

These grooves, or riflings, imparted a spin to the shell and gave it a more even trajectory; but this was not a Krupps invention. German gunmakers in the fourteenth century had first used the idea, but it tended to distort muzzle-loaded lead balls and it only caught on with hunters. Military use of rifled bores (hence 'the rifle'), only became important in the American War of Independence and then only because the American hunters showed that they could outshoot the British who were armed only with smooth-bored muskets. At this time (1780) the first Ferguson breech-loading rifle was employed, though it never overtook conservative muzzle-loading.

The futility of rifle fire led Krupp to invent the first anti-aircraft gun

By boldly employing rifling and breech-loading, Krupps set a standard that others soon came to emulate. Indeed the most powerful muzzle-loaders ever made, the rifled 17.72-inch Armstrong guns (100 tons in weight), only became operational in 1886; the very year that the British government placed orders for its first giant breech-loaders.

Breech-loading guns were ideally suited for mobility but the first combination that led to the tank did not arise until 1916. All the earlier mobile fighting units seemed to be thought of in terms of machine-gun carriers or gun-toting bicycle riders. Yet first steps toward the tank had been made back in Victoria's days. In 1898 Ingleton's automatic track, as fitted to his steam tractor, showed the advantages of an endless driving band. His hinged metal bands were only applied to the large rear wheels to give the tractor a chance to make tight turns at the end of a field. But the lesson was there, just waiting to be learned. This track-band, later developed as the caterpillar track, could cope with rough ground, even soggy ground; and it could drive in places that would defeat mere wheels.

The next step was taken in 1903 when the British Hornsby Company developed its caterpillar-tracked vehicle. This oil-powered tractor was first demonstrated to the British Army in 1903, and its value as a cross-country machine should have been obvious, but horses were still very much in favour and no firm orders were placed. Not even the farmers were overjoyed when Hornsby went on to offer them the first crawler tractor in 1907. Being British he gritted his teeth. Being sensible he then sold his patents to the American Holt Tractor Company who re-developed the machines for farm use and did quite well.

It was not until July 1914 that the Holt tractor was first suggested as a military vehicle. This suggestion came in a letter to Lieutenant Colonel Ernest Swinton. And he acted on the idea. It struck him that if armoured and fitted with a machine-gun, such a tractor might be able to break through barbed-wire obstacles; plough through the waste land of mud and shell holes; surmount trenches; and free the way for the infantry. This plan was considered and taken one step further. If the vehicle was big enough it could carry light artillery as well as machine-guns. The first tank on these lines was commissioned from the Lincoln firm of William Foster in July 1915. By September Foster's had created their first version known as the *Tritton Number 1*. This shed its tracks on trial and went back to the engineers who solved the problem within days. A new, modified vehicle then emerged and the first real tank was born. It was smaller than the first production models hence the name, *Little Willie*, but it came to serve as a valuable training tank for would-be drivers.

The final design approved in April 1916, made provision for two styles of tank. One style, termed a male, would carry two six-pounder guns and four Hotchkiss heavy machine-guns. The other style, the female, would carry four Vickers machine-guns and one Hotchkiss gun. The females were meant to cover the males when they were busy blasting away at machine-gun emplacements or block-houses. These first tanks were huge, slope-fronted craft, with a weight of 28 tons and a length of 26 foot 5 inches. They were easily able to straddle trenches and traverse shell holes. Each used a crew of nine and needed a 105hp engine to reach their 4mph crawling pace. Once the decision to use tanks was made, the army set up its first tank corps in May 1916, but to preserve secrecy for as long as possible the detachment was named the Heavy Section Machine Gun Corps.

The first-ever tank action was ordered as part of the Allied offensive at the Battle of the Somme on 15 September 1916. It was far from an auspicious occasion. Only thirty-two tanks managed to make the front (out of the fifty-nine made) and their intervention failed to turn the tide in favour of the Allies. It was a premature launch; only later uses of the tank met with some success. At the Battle of Cambrai, on 20 November 1917, tanks were used for the first time on ground firm enough for them to display their full powers. 378 tanks helped make the greatest advance in a single day of the whole war. More triumphs came on the second day of the battle and the immense value of the tank was proven. The Germans had already decided to counter with machines of their own. Their first tank of the war, the A7V, was first seen in September 1917. It was truly Teutonic, using two 100hp engines and a crew of eighteen. Six Maxim machine-guns and a 57mm gun made it full of menace. But in practice it played a barely significant role at the front.

Much more effective was the German use of poison gas. This was first prefaced by the use of irritating chemicals at Neuve Chapelle on 27 October 1914. This had little or no impact; so the next step involved the use of Xylyl bromide gas (a tear gas), fired in shells aimed at Russian positions at Bolimov. This first use of gas, on 31 January 1915 was a total failure. The extreme cold froze the liquid gas, and that was that. The first use of a *killer* gas came at Ypres on 22 April 1915. Some 168 tons of chlorine gas was released from thousands of cylinders placed along a 4-mile stretch. The gas broke the Allied defence line, but they rallied and the German breakthrough never took place. Though initially against the use of gas, the British eventually retaliated with a bombardment of Phosgene-filled cylinders at Arras on 9 April 1917. This first British attack was followed by others but the greatest use of gases lay with the German Army, who were also the first to use Mustard gas (12 July 1917, at Ypres). In 1925 the League of Nations organized the first international condemnation of gas and bacteriological warfare. Forty nations signed the Protocol, but the USA and Japan refused. The two future enemies had sent out their first 'No Holds Barred' signal.

It is possible to see the First World War as the first great testing ground of twentieth-century weaponry. The aeroplane had then just reached the stage of dependability. Wireless telegraphy could now be used along with the older line telegraphs and phones. And giant, long-range guns were ready for intimidatory action. Trench warfare led to the development of the first Mills bomb (1915), and the first grenade-launching rifles. The steel shields used in the loopholes of trenches led to the first armour-piercing bullets. Tracer bullets were developed to allow air gunners to correct their line of fire. And the incendiary bullet was concocted to deal with the seemingly invulnerable Zeppelins. To break through barbed wire the *Bangalore Torpedo* was invented, in the form of an extendable set of tubes ending in a 5-foot length stuffed with explosives. The tubes put the user out of danger when the explosive charge blasted a gap in the wire.

To counter the tank, the first pressure-activated land mines were introduced; while at sea the magnetic mine was developed, though it never came to replace the traditional contact mine. The big problem at sea though, lay in the first use of submarines on a large scale.

The value of an undersea vessel had been debated long before the great conflict of 1914. A submersible boat is recorded as being demonstrated in 1575 in the presence of the English monarch. But such a vessel was clearly premature. The primitive construction methods of that time were incapable of turning out a watertight, self-propelled boat able to withstand underwater pressures. By the eighteenth century the advances were just enough to allow American David Bushnell to create the first usable submersible. His egg-shaped, one-man boat was christened the *Turtle*. Inside it, the overworked operator had to hand-crank a helical screw to drive it forward, and with the other arm operate a lever fixed to the steering rudder. At the same time he kept one eye on a depth indicator, and stayed ready to work the ballast pumps and a second helical screw which gave a submersion thrust. And presumably he sang to keep his spirits up. And he would need to, for mounted behind the *Turtle* was a huge mine with a time fuse, just waiting to be armed and released at its target. This was easier planned than executed. The target ship had to be located; a very large screw had to be driven into its timbers; and only then could the mine be released to dangle from a cable fixed to the embedded screw.

The first use of this hair-raising submersible came during the American War of Independence, when it was ordered to destroy Admiral Lord Howe's flagship, HMS *Eagle*. The *Turtle* made its stealthy approach without being detected, but no one had calculated on the difficulty of trying to bore a screw through the thick copper-sheathing that covered the bottom of the flagship. The mine had to be abandoned and it blew up without damaging anything but a shoal of fish. The exercise was never repeated.

It was almost ninety years later before the first successful submerged attack was launched. This time it was made by a multi-crewed vessel, cylindrical in form and made solidly from sheet-iron boiler plates. It was, in fact, based on a discarded boiler. Its propeller shaft ran the whole of its 30-foot length, and was fitted with eight cranks, one for each member of the crew. In smooth waters she could reach a speed of 4mph. She could stay submerged for as long as her air supply lasted. This Confederate secret weapon was first tested in the Mobile river in 1863 and there it sank its first target, a moored flatboat. A torpedo was used to sink the flatboat, but it was very different from the type of torpedo we are all familiar with. A torpedo then was simply a cylinder with an explosive charge and a detonator. These first torpedoes were not self-powered, they were either towed into position, or attached to a long spar and steered to the target by a boat. The method used by Confederate submarine *Hunley* involved a tow rope, a swooping dive under the target, and a sharp impact to fire the detonator and the main charge.

Before the *Hunley* made its first strike against the enemy, its crew were heartened by the news that the first Union ship had been put out of action by a Confederate torpedo. This first successful torpedo attack had been made by the first torpedo boat, a small low-lying iron boat called the *David*. It used a spar torpedo and though partly crippled, was able to make it back to home port. With this example in front of them, the *Hunley* submariners drove their ship towards the Union fleet blockading the port of Charleston. They singled out the thirteen-gun *Housatonic* for attack and on the night of 17 February 1864 the first submarine sortie began. Inside the harbour the blockading fleet felt safe. Admiral Dahgren had learned from the attack by the *David*, and a Confederate deserter had given a detailed account of the submarine boat, so the Admiral had constructed the first anti-submarine boom around his blockade line. But this did not extend to his ships outside the harbour mouth. And the targeted *Housatonic* was one of these. But as the submarine approached, the bright moonlight showed it up as a dark moving shape. The alerted target ship used its engines to back away and the *Hunley*'s torpedo struck the hull prematurely. The *Housatonic* went down at once, but so did the submarine. This first sinking of a warship by a submarine brought fear to the Union blockaders and made them over-cautious; as a result Charleston was able to hold out much longer and did not fall until 17 February 1865. One submarine action had altered the pace of the war.

The *Hunley* killed more Confederate volunteers than Union sailors!

The American Civil War submarines were makeshift vessels born of urgency. Hand propulsion was a temporary solution at best. But the use of an engine in a submarine presented several great problems. The fuel had to be safe in use and all fumes had to be discharged out of the vessel. Electric motors were first tried by Drzewiecki in 1884; but the weighty storage batteries were exhausted too fast. Only short runs were possible. The steam-engine seemed impracticable to many, but the Reverend George Garrett was undeterred by scoffers. He swore by steam power and founded the world's first submarine construction company in September 1878. His Garrett Submarine Navigation and Pneumatophore Company set out to make 'a breathing apparatus, submarine torpedo boat, diving dress and pneumatophore' (the latter was a tank filled with chemicals able to remove carbon dioxide. It was the first closed-circuit breathing apparatus). In his first patent of 8 May 1878 Garrett had toyed with the idea of using a small gas or vapour engine; but his first production model wisely stayed with his first love: glorious steam. In August 1878 his first submarine was completed by J. T. Cochran of the Britannia Iron Works in Birkenhead. But this was more a test-bed than a serious production model. It was used to gain experience in breathing when submerged and in adjusting for buoyancy. The first realistic sea-going submarine was finished on 21 November 1879 and launched on Tuesday 26 November 1879. It was christened *Resurgam* (or *I Shall Arise*), a fitting and optimistic name. The propulsion was now dealt with by a clever method first evolved by Eugene Lamm. Lamm superheated water in a boiler at high pressure. The furnace was then extinguished and the water was forced into insulated reservoir tanks. On opening a throttle-valve part of the stored water would flash to steam at a pressure high enough to drive a steam engine. This system was first used on San Francisco's street cars. Now it had a new use.

After tests in dock it was decided to take *Resurgam* from Merseyside down to the Naval dockyard at Portsmouth. Only there could the Navy be convinced of the value of Garrett's invention. The first submarine journey of length began on 10 December 1879. When Rhyl, (North Wales) was reached repairs were needed. Foul weather now made the trip hazardous and Garrett lingered on in Wales until 24 February 1880. And then he used a steam yacht to tow his submarine out of Rhyl and on to glory. A west-north-west gale hit the convoy within a day. The towing hawser broke and the submarine vanished into the gloom, never to live up to its name. A rumour soon spread that Russian spies had spirited it away.

Garrett's fortunes now looked stygian rather than heavenly, and then he was uplifted by an incredible alliance with the Swedish millionaire Thorsten Nordenfelt. The Nordenfelt Ammunition Company was set on developing both torpedo boats and torpedo-carrying submarines. This urge arose because the company had a stake in the marketing of the Whitehead Torpedo. Though named after Robert Whitehead, this first self-propelled torpedo was first devised by an Austrian Commander Giovanni de Lupis. It first entered production in 1866 and within nine years had doubled its speed to 12 knots. With a range of up to 1,200 yards it was a formidable weapon. Even more so if it could be discharged from a small, fast-moving boat or from one concealed under the sea.

Taking Nordenfelt's gold meant giving up the Garrett name; all the Swedish-made submarines were to be known as Nordenfelts. When the first Nordenfelt was launched in July 1883 it was seen to be the first ever to be fitted with a mounted gun; a Nordenfelt 1.5-inch quick-firer, placed forward of the conning-tower. This craft's first voyage in August 1883 took her from Stockholm to Gothenburg and became the first planned voyage to be completed. But it was some time before anyone offered to buy it; in fact, it was not wanted until the Greeks and Turks went to war in 1886. It was delivered to the Greek harbour of Piraeus on 13 January 1886 and put on trial on 14 March. The Greek navy was thus the first to own a power-driven submarine.

The Turkish government was quick to learn from these events. Back in 1719 Turkey had lusted after a submarine, and one of their naval architects, Ibrahim Effendi, made one to satisfy Sultan Ahmed III. But this first Turkish attempt sank in the Sultan's presence and passions cooled. The Nordenfelt submarine revived the fervour, especially since the Russians were rumoured to have a fleet of *fifty* home-built submarines. So two improved submarines were ordered and the order was marked 'urgent'. The first submarine to reach Turkish soil arrived on 17 May 1886. But it needed to be reassembled and its first launching did not take place until 18 September 1886. Its first trials took place in February 1887 but the first official full-scale trials were postponed until July. This was the first attempt to launch a torpedo from a submerged submarine; it was a miserable failure. The torpedo certainly left the craft, but no one had allowed for the change of trim and the submarine stood on end and plunged to the bottom, stern first.

When the second Nordenfelt was launched the event was marred by the discovery of a spy equipped with a camera that took six pictures on a revolving disk. The lens was disguised as a waistcoat button and the camera was worn strapped to the body and concealed by clothes. The spy never named his masters and he was shot for his audacity. This must be the first case of trial by camera. A trial, naturally, held in camera.

All the Nordenfelt designs suffered from serious flaws. The biggest flaw lay in their inability to stay submerged while stationary. This drawback was corrected by the American designer J. D. Holland, who built his first submarine in 1898, and after a lack of interest from the US Navy, took his plans to England and did a deal with the Vickers Company. In 1901 Vickers landed their first submarine contract with the British Admiralty and delivered the first five craft to the Royal Navy. The German Navy took note and encouraged Krupp to develop a rival design. In 1903 the first Krupps submarine was ready, but oddly enough it was sold to Russia. The design was further refined and from these refinements came the first of the legendary U-Boats (Unterseeboots). The first of this breed, U-1, was launched in 1906.

In 1912 the first diesel engines replaced the earlier crude-oil engines featured in the first batch of U-Boats. The Germans were now ready for submarine warfare on a large scale. And they were the first to prove the killing power of the submarine, when on 22 September 1914 one submarine, U-9, sank three British armoured cruisers in one hour. At the start of the war both the Germans and the Allies relied on diesel engines to power their submarines on the surface, and electric batteries to drive them when submerged. Then, in 1916 Britain reverted to the original Garrett plan of steam propulsion for its first giant K-Class submarines. These craft were meant to keep pace with the Grand Fleet when it set out to search for the German High Seas Fleet. Only a steam turbine could deliver enough power to meet this specification.

The first attempt at a steam design led to the *Nautilus* and the *Swordfish*, two submarines that failed to satisfy. These were then replaced by the K-Class design with a surface speed of 24 knots. In making the first modern submarine driven by steam, Britain seemed to be taking a step backwards. But the value of steam propulsion was once again recognized when the first nuclear-powered submarine was developed. This was first backed by the US Atomic Energy Commission in 1948, who awarded the contract to Westinghouse Electric. The keel of the world's first nuclear submarine was laid on 14 June 1952. Two years later it completed its trials without problems. That first of its class was given the same name as the first British steam-sub of 1916: *Nautilus*. Its nuclear reactor was used to generate steam in a boiler and, since no oxygen was used in the process, this meant that all the oxygen in the submarine could be relied on for breathing. Once an air-purification system had been perfected, this made USS *Nautilus* the first submarine able to stay submerged for an indefinite period. Streamlining the hull and adopting the first teardrop-shaped profile then made a vessel able to reach underwater speeds formerly undreamed of.

The submarine menace brought forth the first depth-charges in 1915. These were canisters filled with explosives and fitted with a detonating mechanism fired by a hydrostatically controlled pistol. A hydrostatic valve was linked to a spring; the spring tension was varied according to the depth at which the submarine was located, and as the charge sank, a point was reached where water pressure would override the restraining spring and the charge would explode. The main disadvantage of these depth charges lay in the need of the attacking ship to pass over the submerged submarine *before* any canisters could be released over its stern, or could be shot over its beams from depth-charge throwers.

It was not until the Second World War that these limits were overcome by the invention of the 'squid'. This was a quarter-deck mounted mortar with three fat barrels able to fire depth-charges way ahead of the ship on a pre-planned pattern. At the same time the 'hedgehog' depth-charge weapon was created. This was a mortar with twenty-four barrels that fired a cluster of small charges, each of 3lb. It was able, like 'squid', to fire ahead but its small charges meant that only a direct hit would sink a U-Boat. The full-sized charges could work without contact; pressure waves from their detonation was enough to cripple or sink a submarine, so the 'hedgehog' fell into decline.

One tricky way to avoid detection yet still hit hard was developed by Italian submarine makers at the end of World War I. They modified a standard torpedo to take two passengers astride it. This could be carried aboard the submarine and launched hours away from the intended target. The passengers would steer their lethal weapon right up to its prey and then abandon it before it wreaked havoc. With one of these manned torpedoes the Italians sank the Austrian Dreadnought *Viribus Unitis* in Pula harbour. This was their first success. The Italians revived this

Italy first used human torpedoes in September 1941

practice in the Second World War, and crippled the battleships *Queen Elizabeth* and *Valiant* in the harbour of Alexandria in December 1941. This first use against them by their former ally, led the British to develop their own manned-torpedo *The Chariot*, which then paid off old scores by sinking the Italian cruiser *Ulpio Traiano* in Palermo harbour in January 1943. These were bold ventures but they bordered on the suicidal. With present-day torpedoes, though, no jockeys need apply. Even attempts at sighting can be forgotten, for cunning acoustic homing devices lead the torpedo to its target with uncanny ease, unless, of course, equally cunning deception devices are both installed and switched on. In that case stalemate and frustration.

• GEE WHIZ! •

The first thermoplastic of the twentieth century was made in 1908. The inventor, Leo Baekeland, called it Oxybenzylmethylenglycolanhydride. But you know it best as Bakelite. He was an ex-Professor of chemistry, who had left his native Belgium to settle down in the USA. His invention was the result of a *failed* experiment. Baekeland was searching for a synthetic shellac with all the same properties as the natural product. At one point he produced a plastic that would mould extremely well but could not be dissolved back into its original state, unlike shellac. So he dropped his initial search and concentrated on finding uses for his new discovery. They were many and lucrative: from cigarette holders to automobile parts, telephones, aircraft propellers, lacquers; the list is an enormous one. In truth his discovery was the first and most important step leading directly to the development of our modern synthetic resin industry and, since we all benefit from that: three cheers for oxybenzylmethylenglycolanhydride!

The first British Admiral of the Fleet to become world famous through an advert, was Sir William James. As a curly haired boy he sat for the painter Sir John Everett Millais RA, who painted him as an angelic creature in a green velvet suit, enraptured by floating soap bubbles. The painting was later bought by Pears' Soap, who used it to advertise their product. It was issued as posters and postcards and was even engraved in black and white for newspapers and magazines. As an advert *Bubbles* became known the world over and *Bubbles* became William James's nickname, one that survived even after he was knighted and rose to the highest rank in the Navy.

The first millionaire to be named after a political party emblem was the son of a New York physician Julius Hammer. Doctor Hammer was a member of the Socialist Labor Party of America in the 1900s, and a very close personal friend of Daniel De Leon, editor of the SLP's *Daily People* and the party's leading spokesman. When his son was born he resisted the temptation to name him Daniel, but felt he had to mark the event with a political gesture. Now, the party's emblem was an uplifted arm holding a hammer, so logically enough (to his way of thinking) the infant was named

Admiral of the Fleet Sir William James, before his promotion!

Armand Hammer. The name never proved to be a handicap. It may even have helped Armand become the first millionaire to advise Lenin on economic strategies involving the USA and the world outside the Kremlin.

The first published proposal for using radium inserts to treat cancer was made by telephone inventor Alexander Graham Bell on 21 July 1903. His letter of that date was printed in the US journal *Science* on 31 July

1903, but was originally sent to Doctor Z. T. Sowers of Washington. In this letter Bell regrets that external rays, whether from X-Rays or radium, have to pass through healthy tissues in order to reach deep-seated cancers. He then states that ' . . . there is no reason why a tiny fragment of radium sealed up in a fine glass tube should not be inserted into the very heart of the cancer, thus acting directly on the diseased material . . .'.

The first territory to be named after a banknote was Dixie. In the 1830s each of the American States issued their own banknotes. Not all of these were held in high esteem, but the notes issued by the Citizen's Bank of Louisiana were always regarded as stable and were willingly accepted. The bank, as an institution of New Orleans, used both French and English on its notes, thus its most popular bill, the 10-dollar value, carried the word DIX on its front along with the figure ten. The bills became dubbed 'Dixes', and when songwriter Dan Emmett wrote some minstrel songs one of them used the line 'I wish I wuz in de land of de Dixes'. This was in 1860. What began as a mere trifle for Bryant's Minstrels caught on as the South began to slide down the slope to conflict. 'De Dixes' soon became twisted into 'Dixie' and the rebels found a song that both heartened them and labelled their precious Southland.

The first Bishop converted to atheism by a penny pamphlet was the Bishop of Arkansas, William Montgomery Brown DD. In 1920 Bishop Brown picked up a slim pamphlet published by the numerically small Socialist Party of Great Britain. It was called *Socialism and Religion* and it argued that religions were a handicap to social progress. The Bishop was devastated by its arguments. No other anti-religious writings, even by scholars, had ever shaken him, but this blunt exposition shook him to the core. So much so that he threw over his Christianity, was expelled from his Church, and then called himself *Episcopus in partibus Bolshevikium et Infidelium*. He was so excited by his new discoveries that he never realized that his 'convertors' were as much opposed to the Bolsheviks as they were to his faith.

The first of the well-loved *Vanity Fair* cartoons was issued on 30 January 1869. The first subject was Benjamin Disraeli and the first artist was Carlo Pellegrini, an ex-soldier of Garibaldi's army. Until his death in January 1889 Carlo contributed chromolithographs week after week, all up to the same superb standards.

The first President of the United States to be impeached was Andrew Johnson. In 1868 a House committee charged him with 'high crimes

'Dizzie', the very first *Vanity Fair* cartoon

and misdemeanors', and on 13 March 1868 he appeared for trial before the whole Senate. The trial ended on 26 May with thirty-five votes cast for conviction and nineteen against. But this vote was one short of the necessary two thirds needed for the conviction to be valid. So he escaped conviction and punishment and retired a few months later, fully and legally acquitted of all charges. Andrew Johnson was also the first (and only) President who never spent a single day in a schoolroom. He only learned to read and write after his marriage to Eliza McCardle.

The first radio receiver using a frog's leg as a detector was created by Professor Lefeuvre in 1921. Lefeuvre of the University of Rennes, knew of Galvani's discovery that the nerve in the muscle of the leg of a common frog is very sensitive to electrical impulses. So he connected a frog's leg up to the output stage of a wireless tuner and fixed a rod and stylus to the end of the leg. When the tuner impulses passed through the leg it contracted and moved the stylus. This stylus, in turn, recorded the signal on the surface of a revolving drum coated in lamp-black. The professor actually received the time-signals from Paris on this strange contraption. Others though, preferred to stick with their headphones and leave the funny tricks to this self-appointed heir of Baron Frankenstein.

The first disk-jockey is usually regarded as Christopher Stone who began his record programmes on BBC Radio. But the fact is that the radio presentation of records interspersed with speech was first heard years before Stone first broadcast. On Tuesday 3 June 1924 Moses Baritz gave the first of his series of hour-long programmes using gramophone records intercut with his comments and elaborations. This first programme dealt with Purcell and other early composers. Baritz is little known as the first DJ mainly because his programmes went out from BBC station 2ZY in Manchester and were never heard nationally. But there is evidence that he was once considered as a serious replacement for Christopher Stone; indeed the Assistant Controller of the BBC even wrote of Baritz '. . . One thing is quite certain, and that is that nobody has a greater knowledge of gramophone records than he . . . ' But Baritz had an image problem. He wore thick, pebble glasses, he was bulky and looked like an unmade bed, and he had a Northern twang to his voice. This made him far too provincial for the suave sophisticates who dominated the BBC at that time, so he never reached the marble Hall of Fame.

The first assassin ever to be filmed was Leon Czolgosz, a young, deluded anarchist who killed US President William McKinley. On 6 September 1901 Czolgosz armed himself with a revolver and went to the Pan-American Exposition in Buffalo. President McKinley was due to visit the Exposition that day and Czolgosz was determined that the President would never leave the city alive. McKinley had agreed to shake hands with the many visitors and Czolgosz joined the reception line and waited for his chance. His revolver was wrapped in a handkerchief to conceal it and this subterfuge worked. He was able to come face to face with McKinley and fire twice through the handkerchief. The President was hit in the abdomen and breastbone, and died eight days later. Could this have been anticipated? The answer came when the Edison company developed the films they had taken at the Exposition. Among the reels they discovered several shots of the assassin calmly waiting to deliver his death blows. But nothing in his demeanour gave any public hint of the violence to come.

The first public advocacy of the use of fingerprints in combating crime was made by Henry Faulds in a letter to the magazine *Nature*, published on 28 October 1880. In it he argues that: 'When bloody finger-marks or impressions on clay, glass, etc., exist, they may lead to the scientific identification of criminals. Already I have had experience in two cases . . . In one case greasy finger-marks revealed who had been drinking some rectified spirit . . . In another case sooty finger-marks of a person climbing a white wall were of great use as negative evidence . . . ' Having made those points (and many more) he then stressed 'The advantage of having, besides their photographs, a nature copy of the for-ever-unchangeable finger-furrows of important criminals.' Alas, no one took much notice at the time, and eight years later Jack the Ripper left his prints all over the place, with complete impunity.

The first detective story involving a murderer caught by fingerprints has usually been listed as Mark Twain's *Pudd'nhead Wilson* of 1894. Earlier still, in 1883, Twain had mentioned thumb printing criminals in his *Life on the Mississippi*. But the first fictional use of fingerprint knowledge appeared in a serial that ran in *Chambers's Journal* from 3 to 17 December 1881. Entitled *The Story of a Thumb-Mark*, it describes how a factory owner is found stabbed and strangled. A bloody knife offers the only clue to the killer, and suspicion falls automatically on the dead man's nephew. The knife is his and he has been in constant conflict with his uncle. But the detective called in has his doubts. Then a bloody thumb-print is noticed on the last, unfinished, letter penned by the dead man. The youth's solicitor sees a way of clearing his client. The police co-operate; and every workman has his thumb-print taken. Each print was then photographed and the glass slides projected together with a slide of the bloody print. An exact match is found and the killer confesses. Happy ending, and drinks all round. The anonymous author of this piece was certainly quick off the mark, since the discussion of criminal fingerprinting had only recently appeared in *Nature* (28 October and 25 November 1880).

The first serious study of the infamous Jack the Ripper murders that chilled London in 1888, was published in Copenhagen, written in *Danish* and named an *Austrian* called Szemeredy, whose only certain crime involved the knifing of a prostitute in *Buenos Aires*. The book was called *Hvem var Jack the Ripper*; the author was Carl Muusmann. An English element seems totally absent!

The first public transmission of still pictures by means of *radio* was made on 30 October 1928 from the Daventry station of the BBC. These pictures could only be received by a special accessory attached to a radio set. This was called a *Fultograph* and it employed a revolving drum which was traversed by a marking stylus activated by radio impulses. The drum was wrapped with a special paper and this recorded the marks of the stylus, which could render a recognizable image, comparable to some of the crude half-tones seen in many an underfunded newspaper. Having said that, you now know why the amazing *Fultograph* never became a best-seller.

The first transmission of speech by means of light rays was demonstrated on 26 April 1880. Alexander Graham Bell, working in conjunction with Sumner Tainter, had created a *Photophone* which was able to work over a distance of 230 yards. When the conditions were right the sender spoke into a speaking tube fixed to a silvered disk. Light falling on this disk was reflected through a projection lens aimed at the distant receiving dish. The dish was a large parabolic device, very much like some of the modern TV dishes, only at its centre was a selenium cell wired into a battery circuit. As the beam of light hit the cell its fluctuations (caused by the voice) altered the resistance of the selenium. Greater or lesser amounts of current flowed through the circuit and energized a set of headphones. These vibrated in time with the fluctuations, and the original speech was heard in the phones. The principle is still in use today.

The first commercial microwave communications equipment was put in use as long ago as 31 March 1931. The devices made by subsidiaries of the International Telephone and Telegraph Corporation, were placed at Calais and Dover and transmitted high-quality signals. These included the first sending of a printed text by facsimile. But all this was more experimental than practical and the first practical use of the microwave technique came on 26 January 1934 when a service was opened between the aerodromes at Lympne, Kent and St Inglevert in the Pas de Calais. This transmitted telephone and teleprinter messages *simultaneously*.

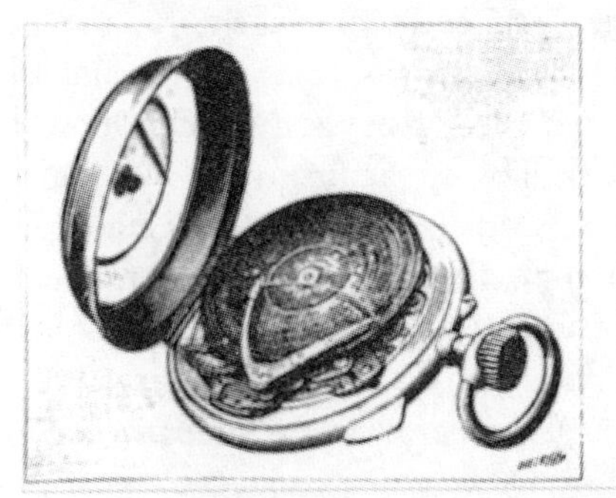
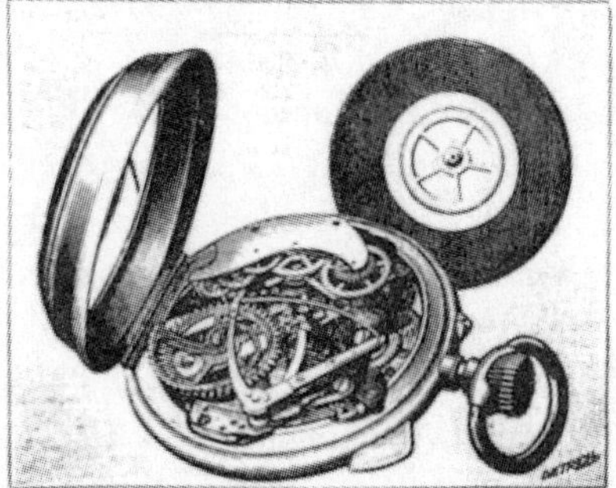

Sivan's talking watch, the mechanism and the phonograph disc, 1895

The first forecast of talking clocks was made in 1877 by Thomas Edison. Having created his first recording machine (it recorded on tinfoil) he gave several ideas about its future. Some were a little extravagant, considering the crudity of his Phonograph, but among them was an idea for a language-teaching machine and a suggestion that timepieces could call out the hours instead of announcing them with chimes. In 1895, after sound recording had been put on a new, practical basis, Mons Sivan of Geneva made the first talking watch. Inside its case was a tiny disk of hardened rubber with 48 voice tracks recorded on it; these tracks covered the 12 hours and the 36 quarters and were synchronized with the watch hands. Provided it was kept wound-up it would dutifully call out the time without being asked. Mons Sivan also made the first talking alarm clock which called out 'Wake Up!' If this was too tame for the buyer, other words, including insults, could be recorded on the clock's disk. Two francs per word and twenty-four hours written notice, please! and the deed was done.

The first automatic dishwasher was invented by Mons Eugène Daugin in 1885. It was designed for a huge restaurant in Paris and used a revolving mechanism fitted with eight artificial hands. The hands seized the dirty dishes and bobbed them up and down in hot water followed by a cold tubbing. Two rotating brushes completed the cleaning and the happy, gleaming plates were then allowed to dry off on a special rack.

The first man to plan a method for signalling to Mars and Venus was the French poet and inventor of the gramophone, Charles Cros. Among his many brilliant ideas was a strange proposal to seek for life on other planets by means of coded signals. His proposal involved the construction of huge concave mirrors able to reflect the light from arc-lamps and direct the resultant beam in the direction of the chosen planet. If there were people out there, and if they had telescopes, then they would notice these bright new lights on the Earth and soon come to see them as an attempt to communicate. Like many of Cros' ideas this remained on paper, and then, in the 1930s Harry Price, famous as a leading Ghost-Hunter, took the idea up and actually drew up the first rough plans for realizing it. He then paid Chance Brothers, the famed lighthouse engineers, to design the reflectors and mechanism. Their blueprint (No. D 903), of 11 December 1929, still exists. A 10-foot in diameter Fresnel lens was to be mounted over a high power arc-light. The whole structure to stand 9-feet high with a tilting arrangement provided by three mechanical jacks. Three of these huge lenses would be needed, so it could not be done cheaply. In the end the costs were far too high for Price, and the grand scheme was laid to rest, so now there is not much point in reviving it – unless, of course, you mistrust NASA; and *some* people do.

The first glasses of Coca-Cola were drunk in 1886. The drink was first formulated by US pharmacist John Pemberton as an improved version of his earlier 'French wine of Coca'. It was born at a time when 'healthy, invigorating' non-alcoholic drinks were in favour in many parts of the USA. Even 'Temperance Bars' flourished, where Root Beers and Sarsaparilla and a dozen more herby and fruity drinks could be had. Only such places or soda fountains, sold Coke, since it was never bottled. This changed in 1922 when a franchise was granted for bottling the refresher. It was this departure, plus a vigorous advertising campaign, that made Coca-Cola a world name. Today half a billion Cokes are sold every day in 180 countries. Yet its first slogan in 1886 was the lame: 'Delicious and refreshing'. Now, would *that* make you want to buy *anything*?

The first photograph to figure in a court action was entered as evidence by the United States Government in San Francisco in May 1858. The action involved land-claims by Joseph-Yves Limantour, a wealthy French capitalist, then living in Mexico. Five years earlier Limantour had laid claim to 17,756 acres of American soil. The area claimed by him was, in fact, the site of the present-day San Francisco! To be fair, he was only claiming half of San Francisco as it then existed. The basis of his case lay in the wording of a series of documents issued by the Mexican Government, the former owners of the territory. These Mexican land grants were held to be valid by a United States Land Commission tribunal in January 1856. Shortly afterwards the US Government decided to lodge an appeal against the tribunal's decision. This followed talk of forged documents and an actual indictment of Limantour on a charge of criminal fraud and perjury. By August 1857 Limantour came up with 'conclusive proofs' that his grant had been properly registered with the national government of Mexico in 1843. When the case came for trial in 1858 the US Attorney brought a series of photographic enlargements into court. One set showed Limantour's papers, the other showed land-grant documents of unquestionable authenticity. When compared side by side it was easy to see that the real government seals on the undisputed papers differed in many details from the seals on the Limantour papers. It was evidence that could not be dismissed and it had a devastating effect on the Limantour claim. The claim was judged as fraudulent on 19 November 1858, but the fraudster had earlier jumped bail and slipped away to Mexico where he lived happily ever after. His views on photographers were never put on paper.

The first underground railway system was opened on 10 January 1863. It was the first part of London's Metropolitan Railway and ran from Farringdon Street to Bishops Road. At first it was handicapped by using both standard gauge and broad gauge (7-foot) lines. It first changed to standard gauge on 1 March 1869. All of its first trains were steam powered. Though the first to open, it was not the first to electrify. The first underground electric service was started by the City and South London Railway on 18 December 1890.

The first spurious banknotes issued by a banknote printer of repute appeared in Portugal in February 1925. They were 500-escudos notes printed by the highly respectable British firm or Waterlow & Sons, who also held the Crown contract for printing the British postage stamps. A Portuguese confidence trickster named Alves Reis cooked up a convoluted story of a syndicate wishing to pump aid into the Portuguese province of Angola. Angola at the time was in financial difficulties, so the idea of a secret rescue-plan was not all that far-fetched. Using faked documents, including a 'letter of authorization' from the Bank of Portugal, Reis sent one of his dupes (Carel Marans) to London to order one-million-pounds-worth of Portuguese notes for use in Angola. They were to be overprinted 'ANGOLA' as soon as they reached Lisbon, it was explained. Waterlow's were taken in by the tale; they printed the notes and handed them over to Marans who in turn gave them to Reis. Reis then promptly opened a bank and sent a second order over to Waterlow's, this time for two-million-pounds-worth of notes. Waterlow's obliged and only learned on 7 December 1925 that the whole episode had been illegal. Reis had been trapped by the fact that he had ordered notes printed with serial numbers *already in use*. Two of these notes came together at an audit, and the game was up.

The first British battleship to be rammed by its own fleet was HMS *Victoria*. The ship was the flagship of Vice-Admiral Sir George Tryon, the man in charge of the Mediterranean Squadron of the Royal Navy. On 22 June 1893 Sir George led his squadron on manoeuvres off the Syrian coast. His fleet was steaming in parallel columns when he gave an order for both columns to turn inwards simultaneously and reverse direction. But the columns were only six cables apart (approx. 1,200 yards) and the turning circles of the ironclads was at least 800 yards. This meant that the warships would be set on an inevitable collision course. Opposite the *Victoria* was Admiral Markham's HMS *Camperdown*, and Markham saw the danger at once and queried the order. But he was told to execute it. Few dared argue with the arrogant Sir George, so Markham went ahead, hoping that Tryon had some last minute master-plan in mind. But no countermanding order came and the *Victoria* and the *Camperdown* steamed towards each other like foes about to grapple. Each ship carried an enormous steel ram on its bows and the inevitable happened. A last minute order from Sir George to change course came far too late and the ram on the *Camperdown* smashed into the flagship's bows. Behind that ram was a weight of 10,600 tons, and the *Victoria*'s armour simply caved in, split and let the sea pour in. The end was horrific. The *Victoria* turned turtle and went down within minutes. 358 lives were lost, and among the dead was Sir George.

The first *Plimsoll Line* was marked on the first British ship registered after 1 January 1876. In its early form it was a simple one inch thick line of at least 12 inches in length, painted longitudinally on each side amidships. Later on it became a series of six lines, each one standing for a different loading condition, such as: Fresh Water, Tropical, Tropical Fresh Water, Summer, Winter, and Winter North Atlantic. No ship could be so loaded that the appropriate line sank below water, without breaking the law. These loading lines were brought in after many ships were lost due to overloading, but it took years of bitter fights in the country and the Commons before action was finally taken. This fight was led by Samuel Plimsoll MP for Derby and in his honour the load lines bear his name. Yet, though he led the struggle against the 'Coffin Ships', the modern load line idea was first formulated by James Hall of Tynemouth in a report of his dated 7 December 1869. Hall was part-owner of a small shipping line and a leading member of the Newcastle-upon-Tyne Chamber of Commerce. He was acutely conscious of the plight of the seamen sent out in overloaded ships and altruistically devoted himself to righting this wrong. Plimsoll heard him speak on the issue in February 1870 and was fired by his words. He embraced the seamen's cause and the rest, is history. But even *earlier* history shows us that the load line was understood, respected and used, long before the Victorian era. The Crusaders to the Holy Land marked their ship's hulls with the sign of the cross. It served a dual role. It identified their faith and at the same time the crossbar marked the limit to which the ship could be loaded. This first line, used in the twelfth century, became the first that passed into law, when the Doge of Venice decreed that the line was compulsory for all Venetian ships. Other countries noted the safety and prosperity of Venetian shipping and used load-lines as well, and the practice lasted until the time of Elizabeth I. Then, with the death of the Hanseatic League, the load-line vanished for centuries, until rediscovered in Newcastle-upon-Tyne.

The first lighthouse dates back to 285 BC when it was listed as one of the Seven Wonders of the World. It stood over 426-feet high and dominated the island of Pharos where it was built. It was said to have been erected at the orders of King Ptolemy II. Its light derived from a huge wood fire burning in an iron cage at its summit. This was tended day and night, until an earthquake toppled this Wonder in 1302.

The first revolving gun-turret on a sea-going ship was fitted to the Union vessel, the *Monitor*, which first sailed on 15 February 1862. The ship was made to lie low in the water, so low that its decks were awash most of the time, but its outstanding feature was its huge rotating turret, 20-feet across and 9-feet high. This armoured structure contained two 11-inch guns and could be revolved to bring its weapons in line with any target in a matter of a minute or so. This brilliant design of John Ericsson's was rushed through to meet the threat of the Confederate Navy warship the *Merrimac*. The *Merrimac* was an armoured gunboat that could strike where she chose and allow cannon balls to bounce off its iron sides. The *Monitor* and the *Merrimac* met in combat on 2 March 1862. This was the first duel between iron warships and it ended with the retreat of the *Merrimac*. Neither ship was badly damaged in the encounter, but the way to New York had been blocked and the Confederates never won sea ascendancy again. After this success the monitor class of warship became the first to be adopted in navies throughout the world. And the revolving turret was used on every class of ship intended for combat.

The *Monitor* in its full glory

The first half-tone process for reproducing photographs was patented by Fox Talbot in 1852. He proposed the use of a gauze or crêpe screen set between the photographic positive and negative when making photo-gravure plates: the earliest form of photographically-created printing plates. He also had the idea of using a glass plate ruled with fine lines in order to break up the continuous tones of a photograph into fine dots. But the first patent to place the ruled screen on a practical basis was not granted until 1865. This (No. 2,954) was the joint work of Edward and James Bullock, who still stayed with the original notion of single-line screens. The first proposal for screens with *cross-lines* was made by F.E. Ives (US). But the first American newspaper photograph using half-tones stuck to the older single-line method. This appeared in the New York *Daily Graphic* on 4 March 1880. The first half-tone subject was nothing grand or noble; it was simply a photograph, by H.J. Newton, of a section of New York's Shantytown! With this inauspicious offering began the great revolution in illustrated newspapers. Until then most pictures were hand-drawn, then cut on to wood-blocks, or transferred to lithographic stones; or they were engraved or etched on to metal plates.

There was an alternative method of printing direct from photographic plates known as the *Collotype* system. This was based on the fact that a chromate gelatine layer becomes case-hardened when exposed to light. If such a layer is exposed under a photographic negative then developed, a printing surface results. This can be used just like a lithographic plate and inked up and printed from. This method thought up by Frenchman Alphonse Louis Poitevin, and first used in 1865, gives maximum fidelity. All the fine details are there, but this very advantage makes it quite unsuitable for the coarse papers used in newspaper production. The *Daily Graphic* example provided a realistic solution to the problem, yet even so, this first half-tone of 1880 did little to alter the standard methods.

It was not until 1893, when Max Levy of Philadelphia introduced his patent process that a real change can be observed. Levy was the first to bring in a commercially viable system of half-tone engraving. He did this by etching fine cross-lines diagonally on glass. The resulting furrows were then filled with a black pigment. Two such glass plates were sealed, face to face, with transparent Canada balsam and a criss-cross pattern of fine lines was thus obtained. His screens could be ruled from as coarse as 55-lines-per-inch, up to an ultra-fine standard of 400-lines-per-inch. Today every newspaper and magazine in the world depends on the half-tone process, yet it was not until January 1904 that the first newspaper appeared fully illustrated by this process. The paper was the *Daily Mirror* printed in London.

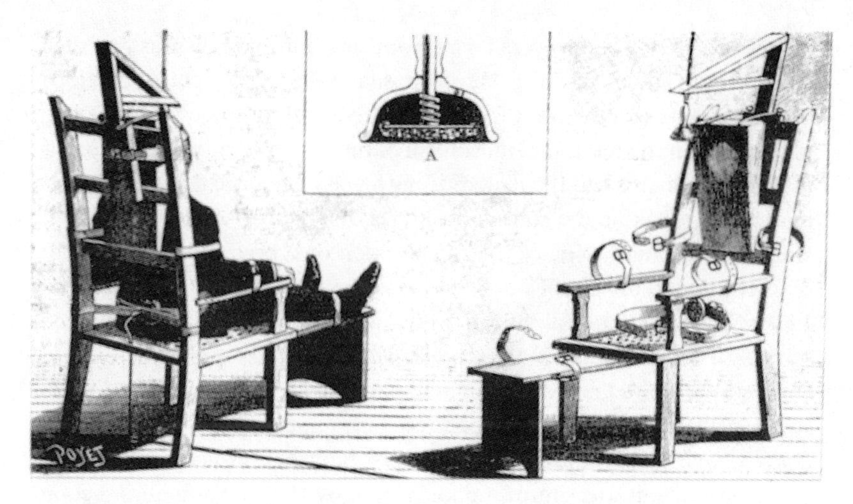

The first electric chair used in New York State to execute William Kemmler

The first electric chair claimed its first victim on 6 August 1890, and first to die in 'The Chair' was condemned murderer William Kemmler. It was claimed that an electric current was swifter and less cruel than hanging, but reports of this execution state that it became '... an awful spectacle, far worse than hanging'. Apparently too weak a charge was used; the first attempt failed and the ritual had to be repeated until the wretch died. The current used for execution was an *alternating* system, not the direct current system pioneered by Thomas Edison. Edison in fact, fought against the introduction of the AC system, which the whole world now uses. Edison's rival George Westinghouse, wisely bought the rights to the first transformer patented by Lucien Gaulard and John Dixon Gibbs in 1881 (No. 4,942). The transformer was the key to the safe commercial use of AC. It made it possible to send high voltages at low currents over long distances; this made for economy because low current power needs less bulk of wire. To use this power at a higher current and lower voltage it was only necessary to pass it through a transformer and tap off the altered voltage from the secondary winding. This was first demonstrated at the Electrical Exhibition at the Westminster Aquarium in 1883. Then in 1884 AC lighting both arc and incandescent was used experimentally on selected stations of the Metropolitan Railway, London.

In 1885 Westinghouse added the Stanley transformer patents to his assets and increased his strength. William A. Stanley had shown the value of his improvements on the Gaulard and Gibbs Transformers by actually building a small power station to serve part of his home town, Great Barrington, Massachusetts. The power plant delivered current at 500 volts and the transformers stepped it down to a safe 50 volts for use.

This first demonstration of AC on a practical scale came at the same time (1885) that Edison was turning down an offer from the German ZDB syndicate to share in *their* efficient transformer. He dogmatically stuck with a system that could only transmit current over a radius of a mile or two from the generator. Longer distances were simply uneconomic. It took twenty years of AC triumphs before Edison, for the first time, accepted that he was wrong.

The world first learned of the discovery of X-Rays on 7 November 1895. The man who released the information was the first to prove the existence of these rays. Wilhelm Konrad Roentgen, Professor of Physics at the University of Wurzburg, had discovered them by accident while using a Crookes tube. These tubes, first made by Geissler of Bonn, were glass vessels with the air sucked out and inert gases injected. Inside these sealed tubes were platinum electrodes and between these metal points an electrical discharge could be induced on connecting an induction coil. As the current passed through the tubes it gave out a soft, coloured light in an attractive stream.

While experimenting with one of these tubes, covered with black cardboard, Roentgen found that some unknown rays were being emitted. These made some substances fluorescent and penetrated paper, cloth, leather and even flesh, blood and muscle. But dense objects like bones halted the rays. Using a screen covered with barium platino-cyanide he was able to show, in a darkened room, the fluorescent effect to other observers. He then placed a hand on a photographic plate, exposed it to the new rays, and on development proved that the rays had given an image of the bone structure with great accuracy. The possibilities for medical diagnosis were obvious to all who witnessed these marvels. For the first time ever doctors were able to obtain a clear sight of complicated bone fractures *before* they decided to operate or use splinting techniques. Bullets, shell fragments and other metal intruders could now be located to a nicety. The most valuable of all the diagnostic discoveries was now at hand. At that time no one appreciated the dangers of excessive radiation. The harsh lessons were to be learned later, but far too late for many.

The first steam turbine was made by Sir Charles Parsons in 1884. It is regarded as one of the most important inventions in the history of mechanical engineering. In his turbine Parsons made the steam act on fourteen rings of moving blades, with each ring placed between a ring of fourteen fixed blades. This produced a continuous ring of steam jets along the whole length of the turbine shaft. By means of graded steps in the diameter of the blades, Parsons increased the working velocity of the steam jets

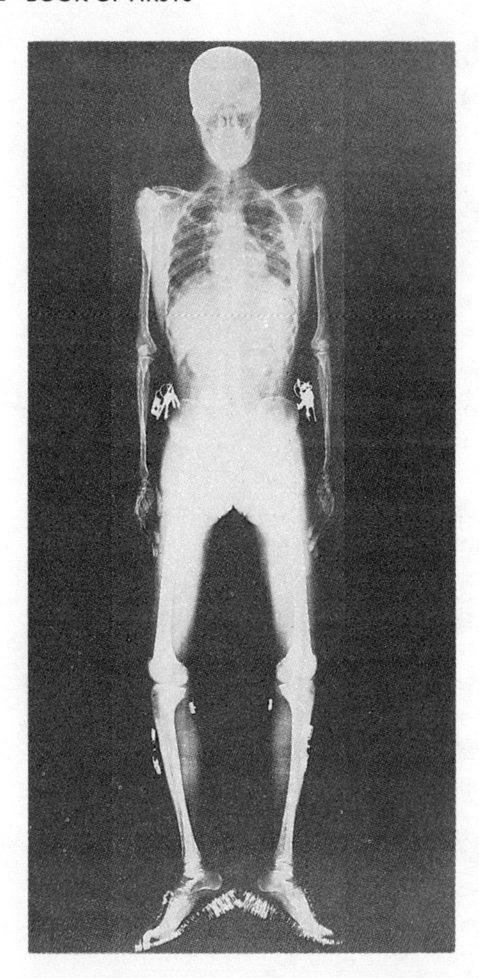

The first X-ray photograph of the human body

in a gradual fashion. This gave maximum efficiency and thrust, without the intervention of a piston-cylinder mechanism. The steam turbine was first used for marine propulsion in 1897 when a Parsons unit was used to power the 100-ton *Turbinia* (first ship named after its prime mover). Parsons turbines developed 2,100 hp and drove three propeller shafts, giving the *Turbinia* a then record speed of 34.5 knots. Before that date the Parsons turbine was first used to drive an electric generator in 1888.

The first hovercraft was developed and patented by Christopher Cockerell on 12 December 1955. He was an electronics engineer with experience of boat-building and his main interest lay in reducing hydrodynamic drag by using *air* as a lubricating medium. His initial experiments were conducted with an industrial blower and a couple of empty tins. From this crude start came a sophisticated design that produced the first full-size hovercraft. This was built by Saunders-Roe (with

Government backing) and as the SR.N1 was launched on 30 May 1959. Her first trials took place at Cowes and then, on 25 July 1959, the fiftieth anniversary of Bleriot's flight, she made the first crossing by hovercraft of the English Channel. But, amazing as this was, the limitations soon became obvious. Cockerell's craft relied on the support given by a cushion of air provided by jets pointing downwards and inwards from the edge of the ship's base. This peripheral jet system did not give enough clearance underneath to cope with large waves. It was best suited to flat, calm seas and how rare they were! Fortunately this very problem had been foreseen by C.H. Latimer-Needham. As soon as he read about the first experiments in 1958 he began to work out plans for a flexible skirt that would contain the air cushion. He reached a satisfactory solution, patented this, then sold the patents to Westland, the controlling body behind Saunders-Roe, in October 1961.

The first Latimer-Needham skirts made by Westland were made of two sheets of rubberized fabric fastened around the edge of the base of the craft. Later on, replaceable fingers were added to the hemline of the skirt. This skirting could raise the craft way above waves and would simply deflect if it hit jetties or rocks. A 4-foot deep skirt fitted to the original SR.N1 allowed it to tackle rough seas and ride over waves as high as 6 to 7 feet. It could negotiate marshlands with gullies (up to 4 foot deep) and cross rough terrain with over 3-foot high prominences. It could now operate at twice the original loading without having to increase power by one iota.

The first service using this new technology was opened on 20 July 1962 by British United Airways. It operated between Wallasey, Cheshire and Rhyl in North Wales, and catered for 24 passengers each trip. Six years later the first Hoverferry service opened between Dover and Boulogne. This was a British Rail undertaking using a Seaspeed SR.N4 ACV, *Princess Margaret*, with an all-up weight of 200 tons and a cruising speed of 60 knots. It could take 254 passengers and 30 vehicles and was the first major commercial service in the world. The flexi-skirt that made this possible has been hailed as a breakthrough comparable in importance to the invention of the pneumatic tyre.

The first saxophone was invented by instrument-maker Adolphe Sax and patented by him in 1846 (Paris). Despite being made in metal it is still a woodwind instrument, though very different from all the other woodwind instruments around at its birth. This is because it has a conical bore which expands at an angle easily three times greater than that of the other conical reeds, such as the oboe, cor anglais and bassoon. This gives it a great mass of vibrating air inside and makes it easy to produce a good volume of sound without effort. But this large interior means that the note holes have to be so broad that only a special keywork can open and close them. Sax took heed of this and designed a clever mechanism that forms the basis of saxophone keywork to this day.

In its first experimental stage the saxophone may have resulted from fitting an ophicleide with a modified clarinet-style mouthpiece, but there is no truth in the often repeated tale that Sax was trying to make the *bass clarinet* overblow at the octave. This is sheer nonsense. Sax was an expert in his field, so he knew perfectly well that the clarinet, with its cylindrical bore, will only overblow at the twelfth. It is this very restriction that gives the clarinet its special tone quality and its great compass. No, Sax wanted a new, powerful voice to enrich the military band and he knew exactly what he was doing when he made this, the first new woodwind instrument of the nineteenth century.

The first commercial solar-motor was demonstrated in Paris on 6 August 1882. It was the brilliant work of engineer Abel Pifre and it comprised a 3.5 metre diameter concave mirror focused on a cylindrical steam boiler. The mirror reflected the sun's rays on to the boiler and when the water boiled the steam generated drove a small engine of about two-fifths horse power. This, in turn drove a Marioni printing-press which turned out 500 copies of a special journal each hour. The solar engine ran for four and a half hours continuously, even though the sky was often overcast.

A printing press driven by solar energy, 1882

The first acceptable lighthouse lens was created by the French physicist Augustin-Jean Fresnel in 1820. He saw that the huge lenses needed would have to depart from the then standard styles, because the thickness of glass involved would cut down the amount of light transmitted. As well as that, the great weight of such lenses would increase the size and complexity of the revolving apparatus. He found the ideal solution in a lens made in steps (lentille à échelons). This could be made of plate glass with a series of concentric, circular grooves, ground, or cast, on its face. The glass between each groove could be so profiled that it formed the marginal segment of a large diameter lens of short focus. The weight of this style of lens would be but a fraction of that of an orthodox lens of the same radius. And it could have its powers increased by the careful placing of prisms around it.

The first lighthouse to be fitted with Fresnel's lenses was the Tour de Cordouan in July 1823. By that time Fresnel had also devised the first concentric-wick lighthouse lamp. This gave a brilliance twenty-five times greater than the single-wick Argand lamps then in use. When combined with his Fresnel lens and prism apparatus, the concentric-wick lamp was able to deliver a beam eight times brighter than that given by any of the reflector systems used on other lighthouses. The first British lighthouse to adopt the Fresnel lenses was the Skerryvore lighthouse in 1844; others throughout the world then followed the French example. Today, the Fresnel lens has found many other uses. It can be found in overhead projectors, on camera viewing-screens and viewfinders, on lightweight hand-magnifiers, on car mirrors and anywhere where a flat, magnifying system is called for. Modern plastics make such lenses cheap, since they can be easily pressed out once a precision master pattern is made.

The first typewriter worthy of the name was made in 1808 by the Italian nobleman Pellegrino Turri. He built it for his blind friend the Countess Carolina Fantoni, in an attempt to make her life easier. The Countess appreciated the machine greatly and the two exchanged letters on an irregular basis over a number of years. In 1908 it was revealed that the State Archives in the Italian town of Reggio Emilia contained sixteen letters and one fifteen-page essay all typed by the Countess on her machine. There were more items at one time, including some sonnets, but these are still missing. From these letters we learn that, at first, the typed impression was made with the use of a sort of carbon paper. Later on this was replaced by an ink applicator. But, since the machine was never preserved, any other details have to be guessed at. What we do not have to guess at, though, is Turri's priority. The evidence of the letters proves the existence of a machine that could instantly record characters on paper and provide these with proper spacing, and regular line sequences. The use of a carbon paper also shows that Turri was the first in Italy to hit on such an idea. He may even have been the first to make this type of paper, since the only known rival in this field is Ralph Wedgwood of England who patented a 'carbonated paper' in 1806.

Turri's machine was a solitary specimen only, so the first typewriter to go into commercial production became the *Skrivekugle* or *Writing Ball*, invented by the Danish pastor Malling Hansen. This machine was first made in 1865 and first put into production in October 1870, by the Jürgens Mekaniske Establissement. The production model was a prime example of precision engineering, made in brass and steel. The letters were fixed on the end of plungers which were set radially in a brass half-ball. When any plunger was pressed it drove down at an angle, hit the paper and on being released was pushed back into place by its spring. All the plungers were so arranged that, if the paper was not moved, they would all hit the same spot. In practice naturally, the paper *was* moved on and the type impressions marched along one, after the other. This machine started off as a large, mysterious, wooden box and ended up as a charming little machine looking like a child's toy. Yet, though it won many medals, and was sold all over the world, it was not to be the pattern that finally determined the office style of typewriting. The winning system was the type-bar method using rows of keys, which settled down at the familiar arrangement of QWERTYUIOP etc. First of this system to be produced commercially was the American Sholes machine. Christopher Latham Sholes of Milwaukee was fortunate in securing backing for his ideas from James Densmore, a small-time manufacturer, that much is certain. The rest of the Sholes saga is so muddled up that it would baffle even Sherlock Holmes to untangle it.

Typewriter historian Mike Adler has taken some fourteen pages in his history to try to make sense of the many contradictory accounts. Though bewildered by the complexities, we can agree that the first patent dates from October 1867 (as the Sholes, Glidden and Soulé machine). The improved machine patent is one of 1871 and the first twenty-five machines were actually constructed in 1871, before work was halted, for some reason or other. At that point George Washington Yost teamed up with them, and super-salesman Yost took the invention to the arms manufacturers Remington and Sons. A deal was struck and on 1 March 1873 the first contract for the making of the improved Sholes typewriter was signed. Production first began in September 1873 and the first machines were placed on sale early in 1874.

Sholes' typewriter [1872]

The first great typewriter was more like a harmonium than anything else

The first author to use the new typewriter was Mark Twain, who also wrote the first testimonial for it in these words, 'GENTLEMEN: Please do not use my name in any way. Please do not even divulge the fact that I own a machine. I have not entirely stopped using the Type-Writer, for the reason that I never could write a letter with it to anybody without receiving a request by return mail that I would not only describe the machine, but state what progress I had made in the use of it, etc., etc. I don't like to write letters, and so I don't want people to know that I own this curiosity-breeding little joker. Yours truly, SAML. L. CLEMENS. Hartford, March 19, 1875.'

The first conviction of a criminal based on fingerprint evidence took place in the Province of Buenos Aires in July 1892. In the small town of Nechochea the two young children of Francesca Rojas were found with their throats cut. Their mother had a superficial wound on her throat and claimed that she had been wounded by her neighbour Velasquez. She went on to claim that she had encountered the man rushing from the murder scene and had been slashed as he passed her. Velasquez was arrested and interrogated but the police were not convinced of his guilt. A request was then sent to the police Bureau of Anthropometric Identification at La Plata, asking for expert help.

In charge of the Bureau was Juan Vucetich, a great believer in the value of fingerprinting in the battle against crime. His enthusiasm and teachings had infected his colleagues and one of them, Inspector Eduardo Alvarez, made a detailed examination of the scene of the crime and took note of some bloody fingerprints on a door jamb. He had the sections with the prints sawn out of the jamb and these, together with prints of both mother and suspect, were sent to Vucetich. The prints were shown to belong to the mother and since she had denied touching her children's bodies, it was certain that she was lying. When confronted with this evidence she broke down, confessed her guilt, was tried and sentenced to imprisonment; since at that date capital punishment was not applied to women in the Argentine.

To Vucetich also goes the credit for starting the world's first fingerprinting bureau. After this great success his bureau identified the bodies of suicides and unknown persons with the aid of fingerprints. This was the first sustained use of the system, but other countries lagged way behind in their appreciation of this new identification aid.

Outside of the Argentine, the first use of fingerprints in a capital case was in Bengal in May 1898. In this case a thief called Kangali Charan was charged with the murder of his former employer Hriday Nath Gosh, whose body was found slashed with a Nepalese Kukri knife. In one of the rifled wooden boxes at the scene was a Bengali almanac with dried blood on its cover; this was sent to the Central Office of the Bengal Police. The Inspector-General of Police for the area was Edward Richard Henry who, in January 1893 started the first criminal record-card system in India that included thumb-prints. In 1895 he began work on a classification system for fingerprints and by 1897 had worked out a system that he offered to the Government of India. On 12 June 1897 a Resolution signed by the Governor-General directed that the system of identification of criminals by finger-impressions should be adopted throughout British India. So the Charan case was the first big test of the claims made by Henry. And in this case a bloody-thumb print on the almanac matched a thumb-print taken when Charan had been previously convicted for theft. This evidence was submitted to the court which perversely threw out the murder charge, then convicted Charan of the theft that accompanied this murder.

In Europe the first use of fingerprints in court involved the murder of a dentist's manservant in Paris. On a window at the murder scene four bloody fingerprints were found. These matched the recorded prints of the convicted criminal, Henri Scheffer of Léon in Brittany. Scheffer was arrested; duly confessed and on 14 March 1903 was given life imprisonment. Two years later the first British capital case using fingerprints was heard at the Old Bailey. This time a double murder was involved and two men stood charged. The killings had taken place at a chandler's shop at 34, Deptford High Street, South London. An elderly man and wife had been battered to death during the course of a robbery and there was little in the way of clues, except for two stocking masks. So the police began to make local enquiries for a *pair* of thugs and learned of a likely couple of brothers named Stratton. In the meanwhile, Sir Melville Macnaghten, Assistant Commissioner of the Metropolitan Police, gave orders for a search to be made for possible fingerprint evidence. Sir Melville had been convinced by Henry (now Chief Commissioner at the Yard) of the value of fingerprints, and his instructions brought to light a very clear thumb-print on a japanned metal cash-box. After a search the brothers were arrested, and fingerprinted and a match was claimed for the right thumb of the elder Stratton. At their trial this match was taken account of by the jury, but it is hard to say how much weight it carried with them. Their verdict was 'Guilty', and on 23 May 1905 the brothers were given the death sentence.

This first major use had been preceded in 1902 by the first use ever in Britain. This pioneer case was one of burglary. A house in Denmark Hill, London had been broken into and some billiard balls had been stolen. A clear print of a thumb was found on a newly painted window-sill and it belonged to old lag Harry Jackson. He was found guilty, other charges were taken into account, and he landed a prison sentence of seven years. Another British first was achieved in 1906 when the New York City Police sent a set of prints by mail to Scotland Yard. The Yard was able to name the owner of the prints as Henry Johnson, a man with seven previous convictions going back to 1894. This was taken into account when the man was later charged and convicted of 'Grand Larceny'. This was the first international co-operation involving fingerprint records.

The first fiction published in weekly parts was *Pickwick Papers* by Charles Dickens. Publication in this form allowed people to spread the cost over a long period, each part being one shilling. Even so, that price was still far too high for the working class of the time, so serialization was of benefit to the middle class only. Some 40,000 of this class were hooked on the idea before part 15 reached the stalls, and other authors, including Ainsworth, Trollope and Thackeray were quick to see that this was a valuable way to have their works issued. Less talented authors thought the same and a flood of cheaper serials, meant for the labouring class, began to roll off the presses. In this way the first 'Penny Dreadfuls' were born. This sudden surge of printed matter was made possible by the introduction of steam printing. By 1840 most big printing shops were steam operated. By 1860 the popular demand for penny serials became so great that one of the leading publishers (Edward Lloyd) even opened up his own paper mill.

The first vacuum flask was devised by French physicist Jules Violle in 1882. But he was not in the least interested in making a flask suitable for picnics or travellers. His aim was to make a laboratory container able to hold liquid gases. He used a double-walled glass vessel with a vacuum between the walls, but never went that one step further along the path to efficiency; he never thought of silvering his glass *internally*. So the first vacuum flask as now used, was a later variation independently arrived at by Sir James Dewar in 1890. He silvered his double-walled vessel both inside and out; this meant that it could retain heat or cold equally well. Using his flasks he succeeded in liquefying hydrogen for the first time, in 1891. In 1902 small versions of his flask were first put on sale by the firm of Reinhold Burger. It came at just the right time. The automobile was beginning to change social habits, and now the Thermos Flask (Burger's trade name) could make country excursions and picnics much more alluring. The choice of hot soups or drinks, or ice cold lemonade was now a simple matter. And in later years the flasks would save many lives by keeping medicines and plasmas in a fresh state.

The first tin cans for preserving foods came into being because Napoleon set up a Society for the Encouragement of Industry. The idea was to encourage new ideas to inspire French industry and prizes were offered for brilliant suggestions. Among the first to win one of these prizes was Nicholas Appert. His speciality lay in his enthusiasm for preserving foodstuffs. He first started by bottling food in champagne bottles, after heating them. He set up the world's first food-bottling plant at Massy, south of Paris. And he sold his products at the first preserved foods shop at 8, rue Boucher, in Paris. The French Navy placed its first official order in 1807, and bottled beans, soups and peas were piped on board ships bound for the Caribbean. In June 1809 Appert published the first book eulogizing the virtues of preserved foods. It was a handy tome with a tongue-aching title *L'art de conserver pendant plusieurs années toutes les substances animales et végétables*.

In 1811 Appert's ideas were taken up in England by Bryan Donkin, a manufacturer with a premises and no work going on in it. Donkin thought in terms of using tinplate containers. He knew all about tinplate, and the industry was able to meet any type of order whether in sheets or cylinders. In 1813 Donkin sent cans of his preserved beef to the Royal family. It met with Royal approval and the first Royal testimonial, which ran: 'I am commanded by the Duke of Kent to inform you that his Royal Highness having yesterday procured the introduction of your patent beef on the Duke of York's table where it was tasted by the Queen, the Prince Regent and several distinguished personages and highly approved'.

Within a year of this praise canned foods were being loaded on board Admiral Ross's ships bound for the Arctic. The first large customer was in fact the Royal Navy and their orders were so huge that it was not until 1830 that canned foods were first put on sale in the shops. Initially the shops only sold sardines, tomatoes and peas. Prices were high and tin-openers non-existent. Recourse had to be made to chisels and hammers at feeding time! Then in 1845 the industry suffered a grievous setback; the Arctic expedition led by Sir John Franklin reported crew deaths that were blamed on bad food. In 1986, for the first time, we learned for certain just what went wrong with the food supply. In 1984 Professor Owen Beattie of the University of Alberta exhumed the bodies of two of the Franklin team. The bodies, of PO John Torrington and AB John Hartnell, were in extremely fine condition, almost life-like. Professor Beattie's autopsy now shows that the men died from *lead poisoning*. Acids in the preserved foods had dissolved the lead solder used in sealing the seams and lids of the tins, and the men were doomed. Rest easy though, modern methods have bypassed these problems. And tinned food can often be safer than fresh food. Contented cats endorse these sentiments every day.

The first vacuum cleaner arose out of a demonstration of a badly thought out invention, intended to remove dust from carpets. This worked, or was supposed to work, by blowing jets of air at the carpet and trusting that the disturbed dust would find its way into a container. Hubert Cecil Booth saw this invention at work in the Empire Music Hall, London, in 1901. He saw the futility of the system and declared that it would be better to suck the damned dust up. He demonstrated this idea to some sceptics by sucking away at a plush restaurant seat, having first put his handkerchief in front of his mouth. Several chokes later he waved the handkerchief towards the unbelievers and showed them the black deposits on the white cloth. Sucking had done the trick. He then worked out the plans for a large, industrial-type machine and patented this in 1901 (No. 17,433). He went for a large machine that could be hired, rather than a domestic machine, because at that time few people had electric power supplies. His machine was so large, in fact, that it was carted around on a four-wheel horse-drawn van. Along with the machine went a portable power unit using petrol, just in case the electric motor could not be plugged in. Hoses, some 800-feet long, were used to take the suction into any room specified by the hiring client.

Booth's Vacuum Cleaner Company gained massive prestige when it was hired to clean part of the blue Coronation carpet in Westminster Abbey. The grimy section lay under the twin thrones and there was no time to haul it outside and clean by conventional techniques. So, in went the writhing hoses and out came the dust of ages – just in time for the Coronation of Edward VII. The King enjoyed this tale of triumph over grime and he ordered a demonstration at the Palace itself. Booth obliged and won two orders for complete outfits; one for the Palace, the other for Windsor Castle.

It soon became the done thing to have one's stately home or Belgravia mansion vacuum cleaned, and Booth even provided transparent sections, so that hostesses could goggle at their dirt as it vanished into the waiting container. Reasonably enough, the demand soon grew for a portable cleaner, and the first was offered by Chapman and Skinner of San Francisco. Their model of 1905 was portable in name only. It weighed over 90lbs and had to be pushed around on a trolley. One had to wait

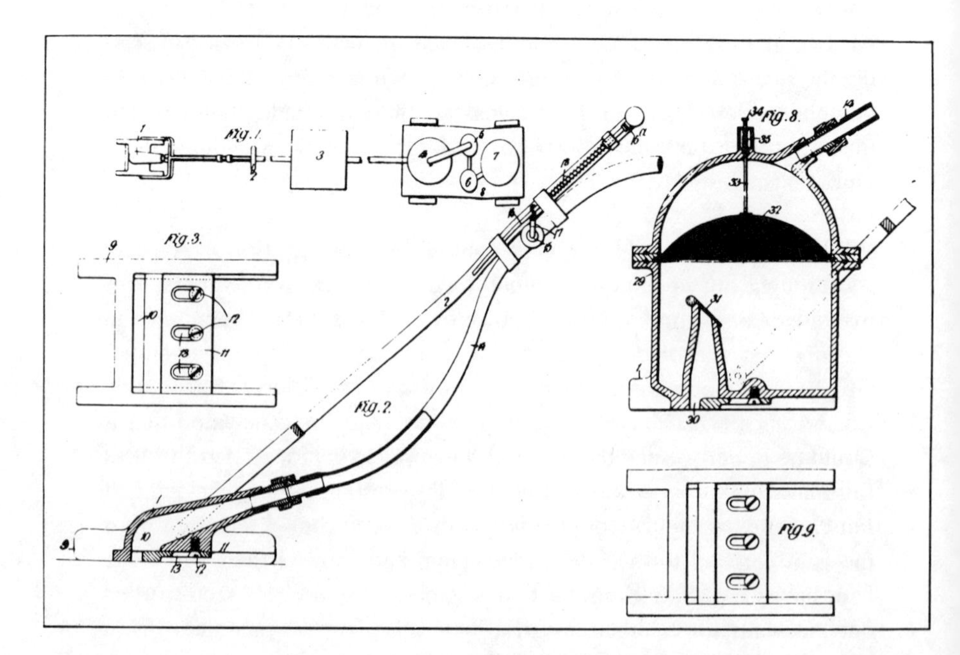

Hubert Cecil Booth's vacuum-cleaner patent

until 1907 before a prototype, light-weight cleaner was devised. At first it was a crude upright design, made from bits and pieces by J. Murray Spangler. He was a low-paid janitor with a department store in Canton, Ohio; and simply lacked the cash to do better, but his ideas were good and sound. So good and so sound, in fact, that a harness-maker from the same town saw the value of his machine and bought the rights without quibble. His name was W.H. Hoover and his first commercial model went on sale in 1908. That first machine priced $70, founded the Hoover empire, and first conquered British carpet dust in 1912. Soon all vacuum cleaners, no matter whose make, became known as Hoovers. This is fitting, since it is one of the few names that can be pronounced by sucking in. Try it.

The first new woodwind instrument of the twentieth century is the Heckelphone. This is sometimes wrongly classed as a bass oboe, but it looks different, sounds different, and has internal dimensions which deviate from the principles used in oboe making. The Heckelphone was invented in 1904 by Wilhelm Heckel, the famous bassoon maker of Biebrich-on-Rhine. He said that the idea was first put in his head by Wagner, who was looking for a more manly, more robust sound than that given by the French bass oboe. The instrument is 4-foot long, with a globe-shaped bell and made in maple. Its bore is conical but very wide, not slender like the oboe family. Its reed is half-way between that of a bassoon and a bass oboe, but the sound it makes is broad and satisfyingly unlike any of the other woodwinds. Its inventor said of its tone, 'It is voluptuously sonorous yet sweet; blooming and rich in harmonics, and so manly and baritone-like that one might be listening to a male voice'. Delius has scored for it in his *First Dance Rhapsody* and Richard Strauss has used it in *Elecktra*. But it was first heard in Strauss' *Salome* in 1905. While its first use in chamber music was by Hindemith in his Op.47. This trio of his, for viola, Heckelphone and piano has been held to be one of his most masterly works.

The first Bloggoscope was created in the 1950s by Bill Blogg. This, I swear, is true. Mr Blogg was then with the Air Survey Branch of the Ordnance Survey and the unit was attempting to use air photographs for full-scale map production. The biggest block to progress was the uncertainty of the aerial camera position at the time of each exposure. Many factors could cause the camera to tilt, and this degree of tilt could vary from one minute to the next. Yet once these tilt angles were known a compensating tilt could be given to the easel of a rectifying enlarger and the printed picture would become accurate.

In his garage Bill Blogg made a prototype *Tilt Finder* by fitting an upright column to a horizontal wooden board. An adjustable carrier on the column carried a contact print from the air camera negative. Above this print was a light source and nine holes allowed light to shine through nine control points on the print. The nine spots of light fell on to nine places marked on a 1:2500 plan of the photographed area. Allowance was made for ground contours by nine adjustable blocks set at the chosen spots, and when the nine spots of light coincided with all the height blocks the exact angle of the taking camera could be read off from scales. Each picture could then be given a setting angle for tilting the enlarger base, and the resulting enlargement would be distortion free. The headaches were over. This brilliant idea was put into use by the Survey; the Bloggoscope was born and Bill was rewarded with £10, since his apparatus qualified as 'a staff suggestion'. But consciences must have been troubled, for the device became a vital aid to map making, so when he retired he was given a further cheque for £1,500 as a sweetener. Still pretty mean, considering that his invention was not superseded until the 1980s, when computers took over the work. Since every map user has benefited from this little-known device, perhaps a pilgrimage to the London Science Museum is called for. For there, in 1989, was placed on display a quarter-scale model of the immortal Bloggoscope.

The Red Cross first came into being in Geneva in 1864, when the Geneva Convention first determined a code of civilized ways for dealing with wounded soldiers in time of war. Before then there were no set rules for caring for the wounded and this meant that nurses and doctors could choose which of the injured they favoured. After the Battle of Solferino in Northern Italy (24 June 1859) the victorious French and Italian forces neglected the beaten Austrian wounded. They even imprisoned the Austrian doctors, who could have looked after their own. One man who witnessed all this and was revolted by the inhumanity was a Swiss businessman Henry Dunant. Dunant was a part-time philanthropist, more concerned with poverty than nursing, but the shock of the miseries of war changed his life. First he organized a band of women from Castiglione, nearest town to the battlefield, and set them to work to nurse the wounded. He overcame their wish to nurse only the French and Italians, by insisting that 'All men are brothers'.

When he returned to Switzerland the nightmare memories failed to fade. To get the horror out of his system he went to Paris, spoke about his experiences and tried to animate people to act to change the wrongs. But, fatalistically, people just shrugged their shoulders and turned away. After three futile years Dunant decided that speech was not enough, so he went into print with his *Memory of Solferino*. His book marks the first dawning of the Red Cross movement. In it he told of the misery he had seen and urged that the nations of the world should come together and agree on humane standards of treatment for the wounded. 'If stopping war is impossible, then we must do what we can to stop some of the horrors of war': this was his central theme.

This book of 1862 tugged at the hearts of all who read it. Letters of support flooded in, and in Switzerland, Gustave Moynier the President of the Welfare Society of Geneva, came to talk with Dunant. Moynier was so impressed that he persuaded his Society to back Dunant's pleas with actions. Four Society members were drafted on to an International Committee and invitations were sent out to politicians, monarchs and army generals. The invitations were to an international conference to be held in Geneva in 1863. Sixteen nations heeded the call; the conference took place and the delegates went back to their homelands with reports and proposals. Now everything turned in favour of Dunant's ideas. A new meeting was quickly called for and in 1864 the Geneva Convention met and agreed that from then on every wounded soldier was to be cared for and treated without regard to his nationality or rank. Those who undertook to deal with the injured were to be given special protection and treated as neutral. To win this protection all helpers, ambulances and hospitals were to be clearly marked with an internationally revered sign. The sign chosen was the red cross on a white background; this was the Swiss flag with reversed colours, in honour of the role played by Geneva. The Red Cross was first seen on the battlefields that very year, during the war between Prussia and Denmark. Both sides used the sign and both sides honoured the agreements.

In later years Moslem countries chose to substitute a red crescent for the cross, while Iran decided to use a red lion and sun, but these are just superficial differences, since the Red Cross standards are agreed to by all. Oddly enough the founding-father Henry Dunant was soon overlooked. He remained forgotten and poverty stricken until 1895 when a newspaper reporter found him living quietly in Heiden. The article that followed reminded the world of its debt to the man. The world took notice; his poverty was relieved and the Nobel Peace Prize was awarded to him in 1901.

The Wankel rotary-engine was first patented in 1958 (No. 791,689) and first used in a motor car in the German NSU *Spider* of 1964. This incredible engine caused a stir in motoring circles when it was first demonstrated. It dispenses with the cylinder and piston combination while sticking with the customary four-stroke cycle. It achieves this by using a rotor shaped like a triangle with convex bulges on its three sides.

This spins in a chamber shaped like two intersecting circles, and this unusual shape allows each corner of the rotor to stay in contact throughout each cycle. A sparking plug ignites the fuel trapped in one segment of space; the driven rotor then takes the exhaust gas to the next segment, where it is emitted; fuel intake and compression is handled by the next segment and so on. This is a very simplified outline of its working; in practice it is more complicated, but the basic value of this rotary motion lies in the bypassing of the many complications involved in turning piston motion into rotary motion. This in turn simplifies controls, gear boxes (you could manage without one) and transmissions. And in theory, since you cut out power losses, this should make for great economy. In reality, though, the Wankel engine is more expensive to make than a comparable piston engine, and there are problems in keeping a gas-tight seal between the rotor tips and the walls of the chamber. Add to that the fact that petrol consumption is higher, power for power, than its piston rivals. This has not deterred the Japanese firm of Mazda who not only turned out several Wankel powered saloon cars but in 1978 introduced their RX7, the world's first Wankel engined sports car.

The last place on Earth to find the father of aviation would seem to be New Zealand, yet many of the inhabitants there swear that the first manned power-flight was made there in 1903. This first-to-fly claimant turns out to be a farmer named Richard Pearce. His farm at Waitohi was also his flying field where Richard tested his ideas and one day actually made a craft from bamboo and steel tubes. With his home-made, lightweight, two cylinder engine well bolted on, this craft was apparently tested while restrained by a hawser to a fence post. Then, one frosty morning, all restraints were off and Richard took to the air. And what date was this? A Mr R. M. Gibson swore that he had seen the event on Easter Monday 1902. Some years later he did concede that it might have been in 1903, but he still stuck to the Easter date. Another source stated that one of Pearce's flights in July 1903 left the plane stuck on top of the 22-feet wide gorse hedge that circled his land. Snow fell heavily before he could winch the plane back into the field, and there it stayed throughout many a blizzard, as a local landmark for weeks on end. Doubt was first cast on this tale, but a recent check of weather reports for the year show that heavy snow did fall in the Waitohi region in July 1903, but none fell in 1904. It is all very strange; but if true, raises many questions. Was it a controlled flight, or a mere hop? Was the field flat, or did the plane start from a raised point? Just how long was the aircraft up aloft? At best, it may well be that Pearce was the first to fly *in New Zealand* for some of his later feats are well documented, and the man could fly.

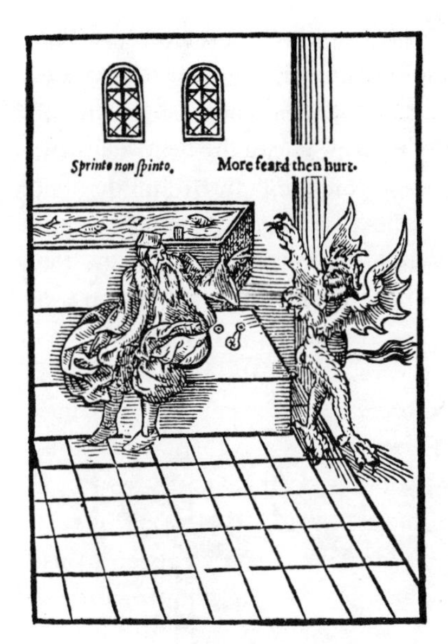

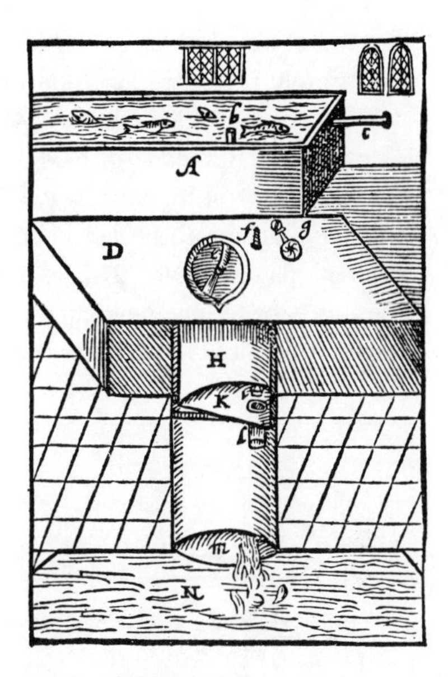

The first water closet was way ahead of its time, 1596

The first water closet was invented and described in full detail by Sir John Harington in 1596. His famous treatise on his invention *The Metamorphosis of Ajax* must count as the first published form of lavatory humour. In his illustration of the cistern he waggishly includes a family of fishes floating around in the water, presumably to make the point that clean water is called for. His description covers all the essentials for a working system; a flush to the basin; a pull-up handle for flushing; an overflow pipe and, of course, the big capacity cistern to keep the bowl reservoir replenished. A working example was first installed in Sir John's own country-seat at Kelston in Somerset. On completion in 1589 it and its designer were both flushed with pride. Yet it did not catch on. This was due more to the problems of disposing of the waste and the difficulties with a water supply, than to prejudice But it should be noted that the monarch of the time, Queen Elizabeth, set a bad example with her haughty attitude to waste disposal. She graciously visited often but stayed at one place only long enough for the privies to get clogged, then she was off to the next retreat, leaving an army of spade wielders to dispose of the problems she had left behind.

The first postal franking machines were introduced in Norway in 1903. The machines were made by the Krag Maskinfabrick of Oslo, but the actual dies for franking were supplied by the Norwegian Post Authority. Each die carried the date, place of origin and postal value.

This was a short-lived service which was discontinued after two years. The country that first introduced postal franking as a permanent feature was New Zealand. In 1904 the New Zealand Post Office approved the franking machine patented (No. 13,360, 1905) by Ernest Moss, of Christchurch, NZ, proprietor of the Automatic Stamping Company. In Britain the first franking machine to be used commercially was made under the Arthur Whitney patents (No. 21,234, 1902) and issued by the GPO to the Prudential Insurance Company head office in August 1922. The earliest known Prudential frank dates the first use as 5 September 1922.

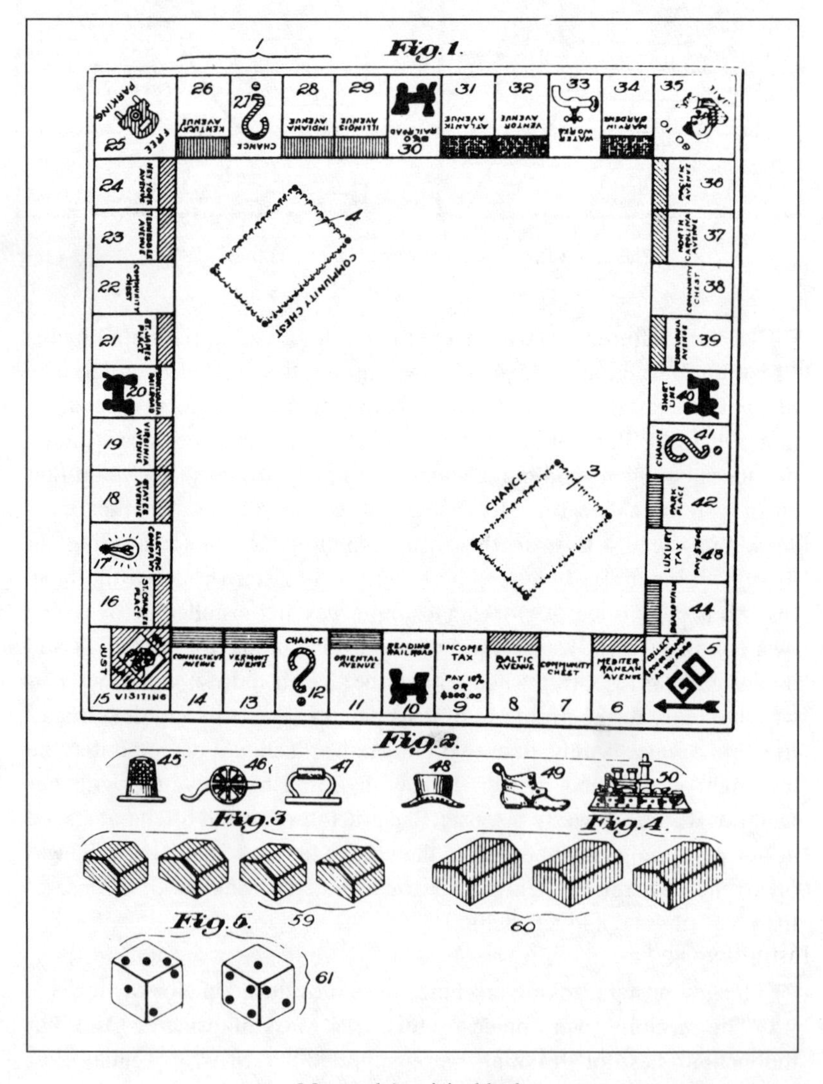

Monopoly's original look

The first game of Monopoly was invented by an out-of-work heating contract salesman, Charles Darrow of Germantown, Pennsylvania. To pass time he sketched the game out on the oilcloth covering his kitchen table. The table-top game was enjoyed by his friends and family. For hours they forgot the Depression (for this was 1933), and became big-time wheeler-dealers, buying and selling real-estate empires, with stacks of paper money. Friends encouraged him to try to market the game, and he had a few sets made up and sold them to department stores in Philadelphia. He then tried to sell the rights to the famous board game manufacturers Parker Brothers of Salem. But they thought the game took too long to play and turned it down. On hearing that Wanamaker's in Philadelphia were buying large stocks for Christmas, Parker reconsidered and agreed to take the game on board. In the first year Parker Brothers sold more than one million sets and it went on to become the largest selling game in history.

The game as played in Britain uses the same basic rules but the board differs. On the original Parker game there is no Fenchurch Street Station, no Oxford Street, in short, none of the familiar names were there on the first printings, for logically enough all the names had to relate to the American experience. The first board design can be seen as the first illustration (No. 453,68) of 1936.

The first printed script meant for the blind was devised and introduced by Valentine Hauy of Paris in 1784. Hauy was the founder of a home for blind children and he created an embossing machine which would press letters of the alphabet on to paper, and create text books for his children. In addition he taught his pupils to print their own books and pamphlets. But the system that gained international importance departed from scripts and used a code of raised dots to stand for each letter. This was the Braille system invented by yet another Parisian, Louis Braille. He was blinded through an accident at the age of three and became a pupil at the Institution des Jeunes Aveugles. The children at the Institution were taught to use the Hauy method, but when Braille became a teacher at the school he set out to replace the raised scripts with patterns. Eventually he evolved the Braille alphabet and advocated it as simpler to learn and faster to read.

He first perfected this system in 1849, but although it was tried at the Institution, and although it was shown to be superior to the old method, it was not adopted in his lifetime. He died in 1852 and it took until 1854 before Braille was first officially adopted by the Institution. In 1868 Braille was first chosen as the standard for Britain and the first Bible was Braille printed by 1890. This project took so long because thirty-seven volumes were needed to complete the work. The first periodical to be printed in Braille was *Progress* issued in 1881. But it was not until 1906 that the first newspaper appeared. This was a special printing of the London *Daily Mail.* Though Braille's name has been attached to various typewriters for the blind it is doubtful if any of them owe much to him. Typewriters for the blind owe most to Pierre Foucauld who made the first in 1839 (*Raphigraphe*). But his first commercially manufactured model the *Clavier Imprimeur* of 1849 departed completely from his original design of ten years earlier, and it was fine enough to win the first gold medal for this class of typewriter, at the Great Exhibition of 1851, in London.

The first silk-screen printing process was initiated by Samuel Simon of Manchester, England in 1907 (No. 756). This was a method of brushing paint through the tiny holes in a textile screen after a stencil had been fixed to the fabric. Paper or card underneath the screen would then be printed with the colour except where protected by the stencil. The fabric (usually bolting silk) was stretched on a frame before fixing the stencil design, while the design itself could be simply painted on with a hard-drying varnish, or made from oiled paper and glued on. The process was made more attractive and precise by the introduction of the first shellac film stencil tissues in 1929. These tissues, invented by Louis d'Autremont, could have the design carefully cut in the hard shellac layer, with all the tiny parts kept in place by the paper backing. Once cut, the shellac side was placed face down on the screen, a hot iron was run over the paper backing and the shellac melted on to the screen, filling the pores wherever needed. At first all the steps in printing by this method had to be made by hand, then in 1917 came the first machine able to squeegee the screens mechanically. This was made by E.A. Owens of California under the trade name *Selectasine*. Today screen-stencils can be formed by photographic methods, making the process fast, reliable and ideal for large-scale posters, though small work can still be handled as well.

The machine that revolutionized the printing industry, the Linotype composer, was first completed in 1855. Its inventor, Ottmar Mergenthaler, was a German living in the United States. It differed from a typesetting machine, since each key released a *mould* of the required letter, figure or shape. When a line of moulds was reached the machine justified the piece, then poured molten type-metal into the hollows and so cast a solid line of type. The machine was first used in 1886 by the *New York Tribune*, and before the advent of film-setting, Linotype came to be used by newspapers throughout the world. Today it still dominates type production in many papers and printers, though its hold on Fleet Street and the world's great newspapers has disappeared for good.

It is far from surprising to learn that the first game of Rugby Football was played at Rugby public school. It originated in the madcap action of a schoolboy, William Webb Ellis. In 1823 Ellis became a right bounder by grabbing the football and running with it towards his opponents' goal. This, of course, made him an absolute rotter in some people's eyes and a spiffing hero with others. Since one was only allowed to *kick* balls, his eccentric change of rules created havoc with ancient traditions. There was a period of deep and soul-searching anxiety on the part of those who cast the rules. Then, one red-letter day in 1841, the first step in creating a new legitimate game was taken. The rules were amended. The sacred leather could now be handled without fear of dire retribution.

From the school, the game was exported to the Universities and in 1860 the first Rugby club was formed. This, the Blackheath Club, had a set of rules which prove that the game has been far from static. They also show that this tough game was at one time near-homicidal. Their first rules state: 'No player may be hacked and held at the same time; hacking above the knee or from behind is unfair. No player can be held or hacked unless he has the ball in his hands. Although it is lawful to hold a player in the scrummage, this does not include attempts to throttle or strangle, which are totally opposed to the principles of the game.' So garroting was outlawed in 1860, but it took until 1871 before hacking was thrown out as well. That year saw the formulation of the first laws of Rugby by the new-born Rugby Union. These laws describe a game somewhat different from the one now enjoyed, since it allowed twenty players a side and held that a goal beat any number of tries. The point-scoring system later replaced this ruling, and in 1874 the number of players was reduced to fifteen.

In 1895 the game split into two jarring factions. The cause of the split lay in the class differences between the public school originators and the new devotees from Yorkshire and Lancashire. The public school people were happy to play for love of the sport; the Northern camp relied on working class talent and it was felt that this talent should be encouraged by payments. Since the two views were irreconcilable a new Northern Union was founded and the break was marked by the adoption of a new set of rules by the secessionists. Northern Union (soon to be known as Rugby League) put the game on a professional basis and reduced the numbers to thirteen-a-side. Many other changes were made and the two games became distinct. Only the shape of the ball remained immutable.

The first synthetic international language was Esperanto invented by Doctor Lazarus Ludwig Zamenhof, a Polish occulist. As a Jew he had direct experience of prejudice. As a Pole he knew of the problems caused by living in an area where German, Russian, Polish and Yiddish were used. He came to believe that a common language would make for brotherhood. But it had to be something new, without roots in past grievances. He made such a language using a twenty-eight letter alphabet, a small vocabulary, a phonetic basis for spelling, and a set of rules that worked without exceptions. He completed his work in 1878 and in 1887 published the first textbook *Lingvo Internacia.* The first national association was founded in France in 1898 and the first in Britain began in 1902. By 1904 the movement was strong enough to stage its first international congress. Other artificial languages then arose in competition, but Esperanto stayed first in the field.

The origin of the first condom is sheathed in mystery, but after considering all the ins and outs of the problem we have to agree that one Doctor Fallopius came first; with a claim that is. And it is quite certain that his claim was given high regard by his contemporaries. Doctor Gabriello Fallopius was professor of anatomy at the University of Padua, in Italy, until his death in 1562. Two years after his death his *Morbo Gallico Liber Absolutismus* was published and from it we learn that the doctor designed a linen condom shaped to the glans and tested it on 1,100 men. The figure seems rather high, but Fallopius was in his post for eleven years, so that reduces it to a manageable experiment. He reports that none of the men using his sheath became infected with syphilis, so we see that his thoughts were of a prophylactic rather than a contraceptive. The birth-control aspect was simply a side benefit.

It was not until 1720 that the condom was first praised as a *contraceptive*, first and foremost. This tribute comes in a poem by White Kennet, son of the Bishop of Peterborough. It rejoices over the liberation brought to young ladies by the all-enveloping condom, which frees them from the 'big belly, and the squalling brat.'

Radium made itself known for the first time when Henri Becquerel resumed one of his experiments with the mineral known as pitchblende. He had broken off his work because the weather turned bad and he needed a good sun to answer this question: did this mineral give off X-rays when bathed in sunlight? In putting his test sample to one side, Becquerel left it on top of a fully wrapped and unexposed photographic plate. He later developed the plate without bothering with the sunshine phase and found on it a foggy image. Was the *cold* sample active? He concluded, rightly, that some part of the mineral was generating its own radioactivity. That part was thought to be uranium. Taking this work one step further, Pierre and Marie Curie then proved that at least one extra substance in the mineral was giving off more radiation than the uranium. After two years of grinding hard work the Curies isolated two new radioactive elements. The lesser-known one was named polonium after Poland, Marie's birthplace, (the first element named after a country) while the second was named radium. This discovery was first made in December 1898, but not fully documented and reported to the Académie des Sciences, Paris, until March 1903. Their monumental work on radiation was rewarded by the Nobel Prize for Physics in 1903, with Becquerel sharing the honour. A further Nobel Prize, for Chemistry, was awarded to Marie Curie in 1911, this honoured her discovery of radium and polonium.

When newly discovered, radium was handled in a quite casual fashion; its dangers were first understood in 1901, when Becquerel suffered a burn on his chest after carrying a tube full of uranium in a waistcoat pocket. These destructive powers were then studied and used to destroy cancerous tissues, where surgery was out of the question. Its dangers became a virtue.

The *Sphairisticke* was first invented by British Army officer Major Walter Clompton Wingfield in December 1873. You no longer find it in the shops or gossiped about; you may even think it looks like an April Fool joke; but without it, the game of Lawn Tennis would not have taken shape. So what was it? It was the first portable tennis court to be taken seriously. It was 20-yards long, 10-yards wide at the ends, and 7-yards wide in the middle. This gave it the shape of an hour glass. The Major's game was played with hollow, indiarubber balls and racquets along the battledore pattern.

This game was tried out at Prince's Club and at Lord's and attracted so much interest that a Lord's committee then drew up the first code of laws for lawn tennis in 1875. And in the same year the All England Croquet Club allocated one of its lawns at Wimbledon to the game. On 25 February 1875, for the first time, Wimbledon and tennis became synonymous. Then the rules were further refined and in 1877 the first championship match was played at Wimbledon. This began on 9 July and ended on 19 July, the first champion being Spencer Gore who claimed the *The Field* silver trophy, worth £25. Further changes were made; the net was lowered; the service line brought closer to the net; and a new scoring system was introduced.

In 1888 the first Lawn Tennis Association was formed by delegates from all the leading clubs in the United Kingdom. This became the body which supervised the game and controlled its laws and regulations. The first great international competition was inaugurated in 1900 as the Davis Cup competition, open to any but restricted to men. First to win the cup was the United States, playing at home at Longwood on a court that, judged by British standards, was appallingly bad. The first international match for women was not seen until 1923. Its trophy was the Wightman Cup, but as the inscription on the trophy shows, this contest was limited to ' . . . Team Championship between Great Britain and the United States'. It was first won by the United States 7–0 in New York 1923. It took another forty years before women players from other countries first had the chance to compete for an international trophy. That trophy is the Federation Cup, first won by the United States at the Queen's Club, London, in 1963 (score 2–1 against Australia). First Commonwealth country to win was the first loser of the first contest, Australia, who won 2–1 in Philadelphia 1964. First European country to win was Czechoslovakia (3–0 against Australia in France).

The first wireless-controlled aerial projectile was invented by Captain Archibald Montgomery Low of the Royal Flying Corps Experimental Works, at Feltham, Middlesex, late in 1916. It took the

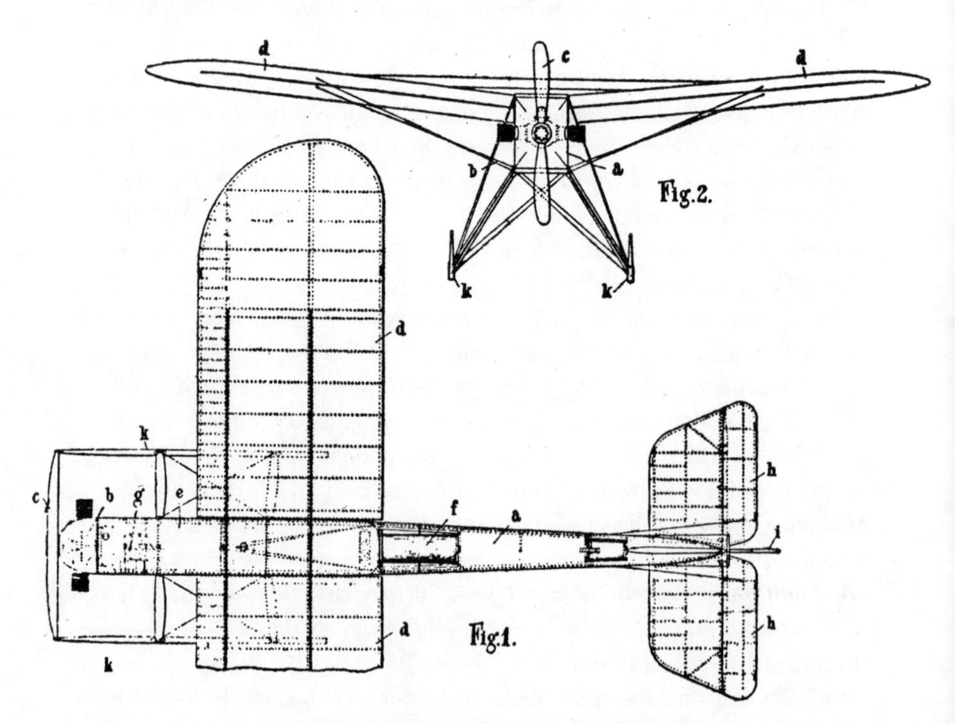

Archie Low's patent for the first unmanned plane

form of a small monoplane with a ski-type undercarriage and a robot mechanism to handle the wing and tail flaps, and rudder. The robot controller was itself under the control of a set of circuits responsive to wireless impulses sent from a ground-based transmitter. Many present-day flying model enthusiasts will be well aware of the way in which ground impulses can guide their models to and fro with ease, and in 1916 this pioneer aircraft established the basic principles still in use.

Low's first full-size Robot Plane made its first flight on 21 March 1917, on Salisbury Plain. When launched Low had the beast under control, but from its spluttering engine he realized that the flight was heading into trouble. Then he saw a group of Army Generals running for their lives pursued by the snarling plane. Fortunately the engine cut out before the brass hats were decapitated and the first trial ended with the plane a total wreck. But the failure was simply *engine failure*; the wireless control and the robot mechanism had been shown to work just as planned. A patent was applied for in 1918, but publication was withheld (by the War Office) until 14 January 1926. This was the first time that the public learned of the device.

Following this trial Low went to work on another missile scheme, but this time he had a steerable rocket in mind. He worked out two alternative control methods for the rocket. The first used a miniaturized version of the unit used in the robot plane, while the other used a controlling wire that fed out behind the rocket as it flew and maintained direct electrical connection with the ground launcher unit. This second method was first employed against Allied shipping in 1942, and since 1950, has been widely used for anti-tank rockets in many a conflict. Though accepted for patent on 8 August 1918, details of this rocket were also withheld by order; the first details were only published on 8 February 1923 (No. 191,409).

Archie Low, later better known to thousands as Professor A.M. Low, tried to claim an award from the Inventions Committee, but all he was given was the thanks of the Court and a firm ruling that as an experimental officer he was excluded from any type of award. Low accepted the ruling with the sad remark that ' . . . invention does not seem to be encouraged in England'. But he did live long enough to see scale models of his robot plane and guided rocket placed in the Imperial War Museum. They were first put on show there on 19 June 1955.

Nylon was first invented by Wallace Hume Carothers in 1934 and first patented 1937 (No. 461,236–7). This American chemist was head of the research team at du Pont's laboratories that was searching for synthetic fibres, which they termed synthetic polymers. They experimented with different quantities of coal, petroleum, air and hydrogen. From these

The launch-day for Nylon

substances they developed hexamethylene diamine, this as a poly-condensed substance has its separate molecules linked together to form extremely long molecular 'chains'. This substance can be extruded through spinnerets to form nylon filaments, which can be drawn out to great lengths, washed and dyed and processed into commercially usable threads. Nylon first went on sale in 1938 and in the same year a rival German product was developed. This was *Perlon*, the discovery of research chemist P. Schlack. Though differing chemically from Nylon, it is such a close relative that it can be freely substituted for it without problems.

The invaluable Xerox copying method was first explored by a man who was cheesed off with being buffeted around by fate. American Chester Carlson was a physicist who lost his research job with Bell telephones in 1930. Of that time he wrote, '. . . I grew up in poverty. During the Depression I decided that making and selling an invention was one of the few possible avenues through which one could rapidly change one's economic status'. Other men thought the same, including the inventor of Monopoly, but in Carlson's case he reasoned that there was a need for a cheap, clean and dry method of copying documents. At the time the only methods around were either the very limited blueprint system, or a wet chemical photographic process, like the *Rectigraph* photocopier (first sold 1907 and patented by G. Bieder). Three brain-boiling years of thought resulted in the electrostatic system using dry powder.

Briefly, the image to be copied was focused on to a paper wrapped around an electrostatically-charged drum, but the charge only stayed on the paper at those places where the original image was dark. A toning powder was then attracted to those charged sections and when fused into position gave a black copy on white paper. This electrostatic principle is one familiar to anyone who has combed their hair and seen pieces of paper or cellophane leap up to the comb and stay stuck to it. Theory was then proved good when on 22 October 1938, Carlson printed his first *Xerographic* image on to waxed paper. This first image simply said, '10–22–38 Astoria'. This bald note marked the start of a multi-billion dollar industry. But it was to take some years before Carlson could persuade anyone to fund development. Some twenty companies refused to buy, or even take shares in, his patent. It was only in 1944 that the non-profit making Battelle Memorial Institute came to his help. An agreement between them gave Carlson the cash to bring his ideas to full maturity. This co-operation led to a deal with Haloid, a small photographic enterprise, in 1947. And in 1959 the first *Xerox 914* photocopier was brought on the market. It came to revolutionize office procedures and ended the days when lengthy transcripts had to be typed over and over again. For years Xerox dominated the field, then the Japanese started to edge in on the market. This, though, initially was a battle of speed and size-reduction facilities. Then in 1973 Canon introduced the first dry copier able to turn out full colour copies, and a completely new phase began. But all these copiers were large office-type machines, quite unsuited for wandering researchers who had spur-of-the-moment needs. Their needs were finally met by the introduction of the first pocket photocopier in 1986. Made by Panasonic, this KXZ 40X measured just over 6 inches in length, by just over 2.5 inches wide. It used a rechargeable battery, which gave it an operating life of twenty minutes between charges.

That strange historic cricket trophy *The Ashes* was first born as a mock obituary. This was printed in the *Sporting Times* after England's defeat by Australia in 1882. The text of this obituary read: 'In Affectionate Remembrance of ENGLISH CRICKET, which died at the Oval on 29th AUGUST, 1882, Deeply lamented by a large circle of friends and acquaintances. R.I.P. N.B. – The body will be cremated and the ashes taken to Australia'. One year later the English team travelled to Australia vowing to 'recover those ashes'. The Australians were trounced and some sporting Australian ladies burned the bails used at the matches, created the first tangible ashes; placed them in an urn and presented them to the victorious visitors. Brought back to Lord's, that urn now stays there in its glass case, regardless of who wins or looses.

The first patent for the refrigeration of foods was issued in Britain to Robert Salmon and Warrel William in 1819 (No. 4,331). But earlier thoughts on these matters had been written down, back in 1755, by William Cullen, chemistry lecturer at Glasgow College. His experiments then, had involved the evaporation of nitrous ether under reduced pressure. This gave him freezing temperatures and some theories to toy with, but led to no practical application until his ideas were adopted by Jacob Perkins in 1834. In that year Perkins made and patented (No. 6,662) a machine operating on Cullen's discoveries. The fate of Perkins' machine is uncertain, but we are certain that the French machines of Charles Tellier were the first to be commercially exploited and used aboard ships to refrigerate meat carcasses for transportation. Charles Tellier's ideas were first patented in 1868 (No. 81,858 Fr) and first used aboard the first specially designed refrigerator ship *Le Frigorifique* in 1877. This ship, a three-masted steamer, took its first cargo of frozen beef, veal and lamb from Rouen to Buenos Aires in 105 days and the meat stayed in first class condition. The vessel of course, has to be the first named after its function.

Many competing freezing systems arose at this time but they were all intended for large-scale preservation. This changed in 1913 when the first household refrigerator, the *Dolmelre,* was placed on sale in Chicago. This was a wooden cabinet design with the refrigerating unit mounted on top, making it ugly to look at, but delightful to know. The first household machines in Britain were late arrivals in 1924, under the name *Frigidaire.* But in Britain there was no great demand for the machines for many years. There was the expense, the need for an electric current (many houses lacked this) and the conflict with established eating habits; together these made the 'Fridge' into a luxury for the few. The big demand came after the Second World War. Secure incomes, increased mobility and an increase in the variety of cheap, frozen foods made refrigeration much more attractive. Frozen foods also prompted the design of the fridge-freezer, and the home-freezer in the 60s and 70s, but by then the frozen-food idea was over thirty years old.

The first frozen-food patent was applied for on 24 August 1925 in the US and granted in the UK on 14 April 1927. It was the work of Clarence Birdseye of Gloucester, Massachusetts who had been impressed by what he had seen on a survey in Labrador between 1912 and 1915. He saw fish caught in fifty below zero weather. He noticed them freezing stiff as soon as they were yanked out of the water. Then he saw them fresh and wholesome when they were thawed out months later.

His patent grew out of these observations; but went much further, since it dealt with the use of fish *fragments* placed in moulds, and frozen as standardized units. This is the first plan for the food that became famous as the frozen fish finger. Fish fingers as such were first introduced by Birds Eye Frozen Foods in the early 1950s in the USA, and brought into Britain in 1955. Other frozen foods had been on sale for many years previously. Birdseye's first firm was set up in 1924, while in Britain S.W. Smedley of Wisbech Canners began selling frozen fruit and vegetables in May 1937. Their first ever sales were of asparagus at 2s 3d per packet.

The first great poem inspired by a photograph (see over) was Henry Wadsworth Longfellow's *Hiawatha*. In 1851 photographer Alexander Hessler of Galena, Illinois, embarked on a search for breathtaking scenery. In August he tramped around the site of the present-day Minneapolis and came upon the dramatic and idyllic 'Falls of Minnehaha'. He took several views of the falls and while developing them in his tent, he met up with a man called George Sumner. Sumner was taken with the pictures and bought two Minnehaha views, one for himself the other for his brother Charles. It was just a passing incident soon forgotten, and at first there was nothing to connect it with an unexpected gift that arrived for Hessler in 1856. The gift was a copy of Henry Longfellow's poems, including his new lengthy poem *Hiawatha*. On the flyleaf was an inscription from the poet himself, reading 'With the author's compliments'. Hessler had never met the poet and stayed puzzled by the gift for almost a year; then he met George Sumner once more, and the mystery was solved. Longfellow had acquired one of Hessler's 'Minnehaha' pictures and taken it with him into the woodlands where he relaxed and dreamed. He found the picture so inspiring that he began to weave a legend around the Indian girl and her life and loves. The result was his epic *Hiawatha*. Without that picture it would never have been born.

The first photograph to inspire a major poem

The first steps towards the laser were taken in 1916 when Albert Einstein set out the theoretical requirements for the stimulated emission of radiation. Years later this inspired a search for a method of putting atomic electromagnetic radiation under full control. Two ways emerged. One used masers, that is microwaves; the other used lasers, that is light waves. Both are known as molecular oscillators. The whole process is quite complicated to explain but it is enough to know that both laser and maser radiation can be sharply concentrated into the form of a beam. The first patent which suggested a practical device to make and use these beams was issued in the USSR in 1959 (No. 123,209); but in its initial form it dates from 18 June 1951. The problem of applying these ideas at that time, lay in the lack of power sources not easily available. News of the first working laser was given in a press release by T.H. Maiman given on 7 July 1960. Maiman had baffled the rest of the seekers by throwing to one side the gases they were all using and substituting *a ruby* as the amplifying medium.

In 1964 Theodore Maiman's ruby laser was first employed by Doctor Freeman to treat lesions of the retina. Then, in 1967, a laser *Light Knife* was developed at the American Bell Laboratories. This beam can cut and cauterize a wound at one and the same time. In the same year a carbon dioxide laser was developed and used to destroy tumours. In 1979 the first 'soft laser' was perfected; this can be used for strained muscles and rheumatic conditions where cuts are not needed. Then followed a series of non-medical applications. The first laser weapon able to destroy aircraft, missiles and even satellites was first tested on 6 September 1985, when the US Air Force destroyed a Titan 1 rocket with the new weapon. And the first holographs were created in 1968. The holograph, or three dimensional picture viewed without external aids, depends on the interference between two beams of light. One beam is reflected on to the photographic emulsion from the scene, while the other goes direct to the film as a reference beam. The basic plan for holography was worked out in 1947 by the Hungarian electrical engineer Dennis Gabor. This was patented in 1952 (No. 685,286) but little could be done with it until the laser came along to give a source of strong, strictly coherent light.

Once the two brilliant ideas came together the pace of movement became sure and certain. In 1984 Visa began using a hologram on their credit cards as an anti-forgery device. Many other credit card companies and banks soon followed this first use. In Australia the Commonwealth Reserve Bank issued the first virtually counterfeit-proof banknote in 1988. This $10 note is made of a special plastic carrying a holograph image which changes colour according to the angle at which the note is seen. 1988 also witnessed the sight of the first holographic postage stamp; an Austrian issue, first used in January. Here, though, security seemed to have played no part in the design, it was more than likely aimed straight at collectors.

The zip fastener, for which be thankful, first came to the mind of a Mr W.L. Judson of Chicago in 1893. His patent (No. 504,037–8 US) describes two metal chains that could be united in an instant by the single action of a slide. It was meant for use on boots and shoes, but although manufactured it came apart too easily to be a great seller. Even the improved version manufactured by the US firm, the Walker Universal Fastener Company, failed to change people's minds. It took a fresh approach to make the zip fastener a best seller. That approach came from a young Swedish engineer living in Hoboken NJ. Gideon

Sunback patented his 'separable fasteners' in 1913 (No. 12,261, 1915) and his patent specification shows the zip that we all know so well. This was the first to become commercially viable; even so it took the outbreak of the First World War to bring about a great influx of orders. Once in the war the US ordered zips for flying jackets, for uniform pockets and even for aeroplane covers. The zipper habit was first created not among the tea cups of the salons but among the grime and grease of warfare. In Britain the habit took off much more slowly. It is said that only the giant working model of a zip shown at the Wembley Empire Exhibition of 1924, made people think again. True or not, by 1927 sports clothes fitted with zippers were on sale in Britain and were bought without qualms. There were some fears however, when they were first used on men's trousers in 1935. These were unfounded. The casualty rate is so low that the hospitals have given up keeping statistics.

The ubiquitous ball-point pen, the Biro, was first invented by Hungarian journalist Laszlo Biro in 1938 (No. 498,997). He stated that the idea of a pen using a *quick drying* ink came to him in a print shop. He saw how fast their inks dried and decided to use something along the same lines in a new type of pen. An ordinary nib would clog up with a thick ink, so he worked on the idea of a tiny ball-bearing revolving in a housing full of this ink. In 1940 he fled from the Germans, first to Paris then to South America, and there he took out a new patent in 1943 and made the first commercial models. Rights to these were bought by the British Government on noting that they did not leak or blot; they saw them as ideal for aircrew use. In Buenos Aires the pens were first sold by the Eterpen Company in early 1945. But Biro had forgotten to patent the pen in the US and so lost the giant market there. In the US it was boosted as 'The first pen to write underwater'. Thousands must have had this strange need since, when the US pen was first launched, some 10,000 sold at the launch-pad in Gimbel's of New York. This was on 29 October 1945. In Britain the civilian version was first sold by the Miles-Martin Pen Company at Christmas 1945. The great defect of all these early models was that the head with its revolving ball was fixed to the body. If the ball failed to work then the whole pen had to be returned to the makers for replacement. All these early pens were quite expensive and it was not until 1953 that the first cheap ball-point came on sale. This was the French Bic pen that became cheap enough to throw away without tears or regrets. The Biro certainly came as a boon to many but it had the unfortunate side-effect of creating a new army of people with unrecognizable scrawls. Previously this privilege had been reserved for doctors of medicine.

The first banknotes to carry wording in seven different languages were issued in 1919 in Soviet Russia. Several denominations were involved, but taking the 5,000 rouble note as typical we find the slogan 'WORKERS OF THE WORLD, UNITE!' in English; in Italian as 'PROLETARI DI TUTTI PAESI, UNITEVI!'; in French as 'PROLETAIRES DE TOUS LES PAYS, UNISSEZ-VOUS!'; in German as 'PROLETARIER ALLER LANDER, VEREINIGT EUCH!', with the same words repeated in Chinese, Arabic and Russian. The notes quickly became known as 'Babel Notes', after the mythical events that first led to the confusion of many tongues.

The first *Catseye* reflector road-studs were first thought up by Percy Shaw of Halifax, England. The idea came to him while motoring along a foggy stretch of road in Yorkshire. He was veering towards the edge of the road and possible disaster, when his lights picked out the shining eyes of a cat and caused him to alter course. That experience in 1933 started his experiments aimed at making artificial catseyes for road use. He finally made a cast-iron container which held two special reflectors in a rubber housing. The reflectors were small tubes silvered on the inside and fitted with a convex lens at the open end. When sunk into the road surface they would be kept in good reflective order by the passage of cars, since every downward pressure on the housing would pass the rubber over the lenses and keep them clean.

The first batch of Shaw's reflectors was used at an accident blackspot in April 1934. But there were other reflecting devices being offered in competition and it was not until 1937 that the catseye was given the chance to prove its superiority. In that year the Ministry of Transport allotted a 5-mile stretch of road for a first major trial of ten different types of reflecting studs. At the end of two years only the catseyes were still working, making it the first choice for safe roads when the war-time blackout was imposed. Shaw's patent of 1936 (No. 457,536) made him very rich but he still stayed put in his drab house, with his only luxury being four television sets! In 1965 he was awarded the OBE for his very visible contribution to safe roads.

The first brassière, as we know it, was patented by Mary P. Jacob in 1914 (No. 1,115,674 USA). The patent drawing shows a cloth lay-out pattern with breast-covering support-sections and back and shoulder straps; clearly the basic pattern still behind most present day bras. There were of course, far earlier methods of supporting the breasts, but these fail to qualify as brassières, since they were either bracing straps or simply the shaped tops of corsets.

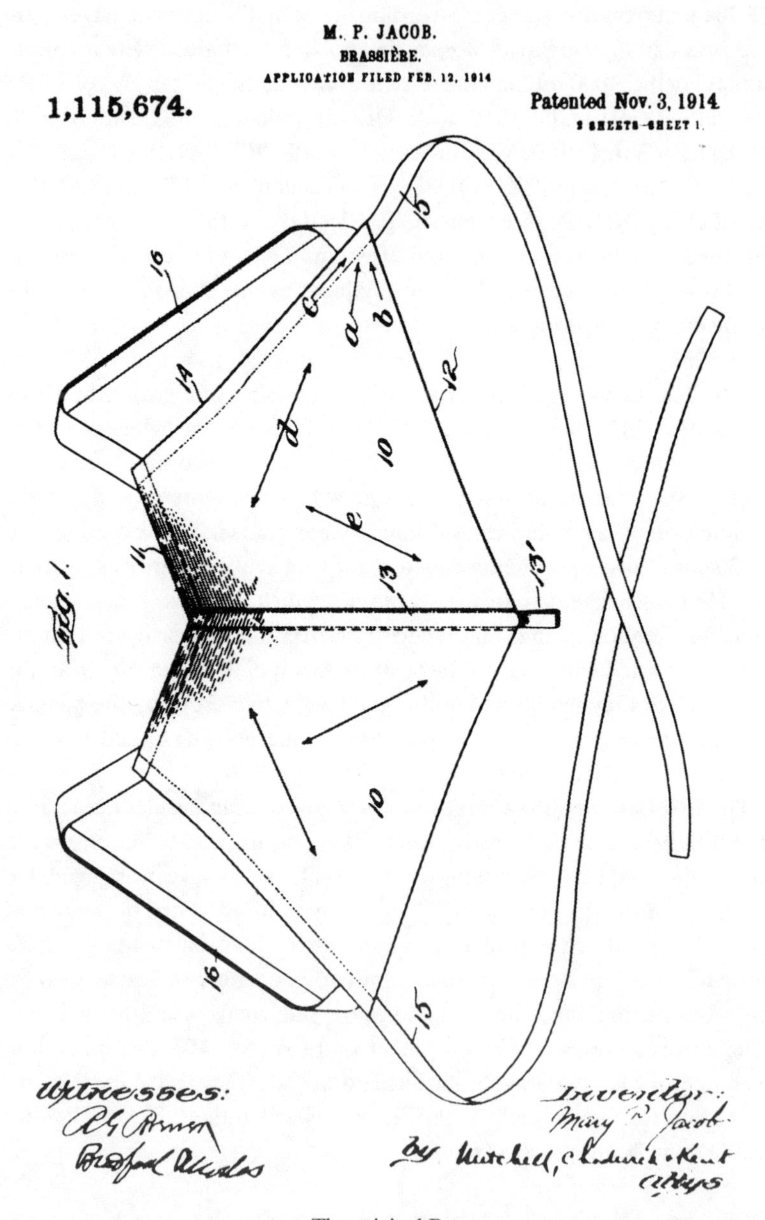

The original Bra

DDT was first synthesized by Othmar Zeidler in 1874. But Zeidler never knew it by that name. He preferred the much more convenient label: Dichlorodiphenyltrichloroethane. He also had no idea of the insecticidal properties of his creation. We owe both the handy name and the discovery of its value to Swiss chemist Paul Muller of Geigy. Once he had shown it to be a potent insecticide in 1939, he sat on the knowledge for a while, then in 1942, he sent full details to the Allies through the British Legation in Geneva. DDT was then used extensively in war zones for delousing and related relief operations. Muller was given the Nobel Prize for Medicine in 1948 in honour of his discovery. It was only much later that the highly polluting properties of DDT were fully recognized. This resulted in a ban on its use in many countries.

The first authorized adhesive postage stamps were issued on 6 May 1840 by the General Post Office in Britain. They were the famous Penny Blacks. Two days later came the first much prettier, Twopenny Blues. The first public suggestion of using a prepaid adhesive label came from Rowland Hill in 1837; but before that, in 1834, James Chalmers of Dundee had not only put forward the same idea but had actually printed samples of such stamps at his printing-works. In December 1837 he submitted examples of his work to the special Parliamentary Committee set up to consider Rowland Hill's idea of a Penny Post. Some of his later essays dated 10 February 1838 survive, proving that he was the first to issue postage stamps, even if they were unauthorized. And it is well worth remembering that Hill's first choice was for an elaborately designed envelope or wrapper carrying the postal price. The adhesive stamp was more of an afterthought with him. Before Chalmers' time, postal payment was acknowledged by franking the envelopes with a hand-stamp. This system was first introduced by William Dockwra for his London Penny Post on 1 April 1680. But after the issue of the first British adhesives all countries saw the value of the idea.

But the first country to follow Britain's lead was not one of the great European powers, and not even the United States. That honour goes to Brazil who issued their first adhesive postage stamps in 1843. These were utilitarian in the extreme; just a figure of value in an oval and no writing of any sort. The use of pre-paid post stationery kept pace with the issue of stamps but the first pre-paid postcard did not arrive until October 1869. First country to come up with the postcard was Austria. Their cream-coloured *Correspondenz-Karte* carried a primrose yellow two kreuzer stamp printed on the card. The example was quickly followed elsewhere and led in 1872 to the first *picture* postcards. These came from Locher's of Zurich and showed local views. This growing world use of postal stationery, postcards and postage stamps, inevitably meant some type of international agreement on their uses *between* countries, so the first body to preside over these matters was set up in 1874 as the Universal Postal Union. The first edict of the UPU stated that all stamps had to bear the name of the issuing country. The sole exception was Britain, in token of the fact that she had been the first in the field. This rule still applies.

The first dental drill is described in Pierre Fauchard's book on dentistry *Le Chirugien-Dentiste* published in 1728. This was hand powered by twisting it back and forth between the fingers. The first power-driven dental drill is described in George Fellows Harrington's patent of 1864 (No. 1,017). This is meant to be driven by ' . . . clockwork contained within a hollow metal box, case or holder of convenient size to be held in the hand of the operator. When the spring is wound up the clockwork immediately commences to revolve and thus imparts rotary motion to the tool'. This style of drill was superseded late in 1874 by the first electrically powered drill; an invention of George Green of Kalamazoo, Michigan. Battery-power had to be drawn on to use this drill since mains power did not exist.

The first paper money was used during the reign of the Chinese Emperor Yung Hue of the T'ang dynasty round about 650 AD. None of these notes have survived but they are shown in early illustrations. In Europe the first paper money is said to have been issued by James I of Catalonia and Aragon in 1250, but in this case there are no illustrations to back up the claim. But we do know that the first *provable* paper money issue dates from 1574 during the siege of Leyden. This siege was just one of the horrors of the eighty-year-long war between Spain and the Netherlands. Since all metals were needed for arms the Burgomaster Pieter Andriaanszoon ordered the coin presses to be fed with paper squares made from the compressed pages of books. This paper took up the familiar designs and was treated as equal to the metal money it had to replace. The first *banknotes* appeared in the next century, when the Swedish Bank of Stockholm ran into trouble in 1660. At the time the currency had been depreciated, so to prevent a panic and a run on the bank, the bank founder, Johan Palmstruch, bought time by issuing notes of credit or *Kreditvsedlar*, as they say in Sweden. Palmstruch's position as the inventor of the first banknote is not in dispute, anywhere, which makes a change.

The first lift was installed at the Palace of Versailles in 1743. It was placed on the outside of the building for use by the monarch. It was supposed to give the King a quick journey to his mistress's apartments in the floor above him. Perhaps stairs would have taken up too much of Louis XV's amorous energy. The first passenger lift not involved in assignations was sited at the Regent's Park Coliseum in 1829. It took from six to eight people at a time to the viewing gallery of the giant panorama set high up in the dome. By 1848 this lift or elevator could take from ten to twelve passengers at one time. How it worked was never made plain. The first use of an elevator in a department store dates from 23 March 1857. The place was the Haughtwout Company's five-storey store on Broadway, NY. The lift was supplied by Elisha Graves Otis, the first manufacturer of passenger elevators.

The first hydraulic lift was a French invention first used at the Paris Exhibition of 1867 and later fitted into the Eiffel Tower. It was twenty times faster than the American Otis pattern elevators, but needed very deep foundations to house the huge hydraulic pistons and cylinders.

The first electric passenger lifts were installed at the lofty observation tower of the Mannheim Industrial Exhibition of 1880. These were made by Siemens and Halske and reached a height of 72 feet in eleven seconds. The development of high-speed passenger lifts altered ideas of town planning. Now it became possible to think of building higher and higher, even to skyscraper levels. Without lifts such constructions would never even have reached the drawing board.

Fluorescent lighting came about due to the need to economize on electricity in large commercial undertakings and in public places. The standard filament bulb is not a very efficient user of electricity, but until 1901 there were no alternatives. A search was made for a tube that would dispense with the heated wire of the bulb and generate light by discharging an electric current through a gas. First to succeed with this principle was the Cooper-Hewitt Company which produced a mercury vapour discharge tube in Britain in 1901. Their light was bluish and not greatly popular. It tended to make people look rather ghoulish. The next step was taken in France when George Claude brought out his first discharge tubes filled with neon gas, in 1910. These were used to light the Grand Palais in December of that year; but though not so ugly as the mercury tubes, their light was still limited to a narrow band and gave a reddish-yellow look to people and objects. There were other disadvantages. They were bulky and quite expensive and this ruled them out for domestic use. As street lights they were passable. These limitations threw researchers back to the mercury vapour tubes; only this time a new principle was applied. It was known that some substances, the phosphors, would emit light if bombarded with ultra-violet rays. Now, the mercury vapour discharge gives off *unwanted* ultra-violet, since these rays can be harmful. But by internally coating the glass tubes with phosphor compounds, these rays can be shielded and put to work. A fluorescent light then results. The first tubes made on these new principles were the hot cathode designs made under the George Elmer Inman patents of 1933 (No. 390,384). The first practical use of fluorescent lighting was made at the Centenary banquet of the US Patent Office held in Washington on 23 November 1936. Commercial production came later in April 1938, when the first tubes were marketed by both GEC and Westinghouse in the USA. Seven different colours were then on offer in three different sizes.

The first modern family Holiday Camp started up in 1906 as a centre for political up-lift. Based at Caister-on-Sea, *Dodd's Socialist Holiday Camp* was intended to give working class families a chance to enrich both body and spirit. It was the idea of J. Fletcher-Dodd, a member of the Independent Labour Party, and a believer in 'the healthy life'. This meant no booze, plenty of exercise and sports, and a hardy life under canvas. Rattling good debates were organized, as well as edifying lectures, and the whole feel of the site must have been a cross between an army camp and a pioneer wagon-train's halting place. The very thought has its own strange charm, for some eccentrics at least!

The first and only British Prime Minister to be assassinated was the Right Honourable Spencer Perceval, who was shot dead in the lobby of the House of Commons on 11 May 1812. His killer was John Bellingham, a man who had some imaginary grievance against the state. For eight years, he claimed to have fought for justice and his rights without being given a fair hearing; but when he had his chance in Court, his presentation was so confused, that to this day, no one is sure just what his problems were. He was hanged for the murder.

There is no truth in the yarn that the PM's life could have been saved if only the authorities had heeded a warning given in a dream. And there is no truth in the claim that details of this dream were first printed in *The Times* just days after the shooting. This dream is said to have visited a Cornish mining engineer eight days before the fatal day. This Cornishman John Williams, dreamed of a small man in a blue coat and white waistcoat being shot by a larger man wearing a snuff-coloured coat with yellow metal buttons. The event took place in the House of Commons. On the surface this looks convincing. His description of the place and the clothes worn by the two men was accurate. Unfortunately, however, the first published account of this dream did not appear in print until 28 August 1828, over *sixteen* years after the events. This account in *The Times* was followed four years later by the first public statement from John Williams himself. This was circulated as a two-page leaflet and repeated in Walpole's *Life of Perceval.* The account of 1828 clashes with the account of 1832 and differs from yet another account which he is said to have narrated in 1815. At best we can say that some type of nightmare terrified Mr Williams, but its prophetic value rates as nil. No written record made at the time, or even days later, ever existed.

The first rubber factory was established in Vienna in 1811 but its product was treacherous. In cold weather it was docile but with the coming of summer the heat made it sticky and quite malodorous. In Glasgow, Charles Macintosh took an interest in this anti-social material and tried to tame it by dissolving it in some oily white naptha. The idea worked and gave him a varnish-like fluid that could be spread out to give a thin sheet as soon as it dried. He next smeared this wet naptha-rubber over a length of woollen cloth; did the same to a similar piece; pressed them both together and invented the first Macintosh material. But there were problems with tailoring the stiffish cloth and these were not resolved until James Symes discovered a finer solvent. Macintosh snapped up the improvement and patented the first commercial *Macintosh* material on 17 June 1823. Thus began the great waterproof garment industry, with the first important commission coming from Sir John Franklin's Arctic expedition of 1824. There were still problems with hot weather, and still complaints about the strange odour, but nothing beat them in stormy weather. One man who tried to further improve on the Macintosh method was the tragic figure of Charles Goodyear. From a well-placed American family, going back to the founding fathers, he seemed to have all in his favour. But the slump of 1830 damaged his business and Charles tried to avoid bankruptcy with long-term notes and promises. These were to burden him for the rest of his life. Heavy with this burden he dreamed that he could alter his fate by solving the great rubber problem. He kneaded raw rubber with every type of material in an attempt to find one that would stabilize it. Eventually he treated some rubber with impure nitric acid and found that the stickiness vanished. But it was the sulphuric impurities that did the trick. Once this was understood Goodyear had a method that would work in part, but only on the surface of the material. Still it was better than nothing and it was patented.

The real breakthrough came with the accidental discovery of vulcanisation. A piece of his rubber treated with sulphur became charred by a hot stove. At once he saw that a controlled heat-treatment would give him the curing method he had chased after for so long. He now sat on a treatment worth millions and no one was willing to back him. That first great discovery of his was made in 1839. But it was three years more, before Goodyear began to claw his way out of the poverty-trap. With accumulated debts of $35,000 the task was onerous and although his

Charles Goodyear, the brain behind the rubber revolution

family began to eat well once more, he never reaped the full rewards for his years of labour.

Goodyear's work gave us the first rubber fit for industrial use. New problems then surfaced. The supply of the raw materials depended on wild trees whose milky saps varied greatly in quality. The finest grades came from the hevea trees in Brazil and these were naturally the first choice of every manufacturer. But the increasing demand pushed the prices up and up.

A dilemma of major stature faced an industry chasing more markets. A fast answer was called for, and it came through chemistry. In 1899 American chemist Arthur Marks discovered the first method of reclaiming used rubber. He marinated all the scrap for twenty hours in a weak solution of caustic soda kept at a high temperature. This eliminated any cotton fabrics and free sulphur and made the rubber plastic once more. Overnight the industry had a method of keeping its costs down and had won time to look for new ways of increasing the quantity and quality of the wild rubber supplies.

Once again Arthur Marks came to the rescue. This time he hired an old classmate of his to join him and search for a magic quality-raiser. If the cheap, inferior gums could be raised up to match the top-grade Para gums, then his firm alone could save $20,000 each day. His new left-hand man, George Oenslager, set to work by mixing chemical after chemical in with the raw rubber. Within six months, following hundreds of frustrating experiments, Oenslager found the magic powder, Thoicarbanilide. When mixed in with the cheapest gums it created a rubber that was superior to that resulting from the expensive Paras. Arthur Marks' firm, the Diamond Rubber Company, first used the new process on 22 June 1906.

The rest of the industry joined in within a year and the cost of tyres dropped, while their useful lives increased over five-fold. That first use of analine determined the future fate of the motor tyre and set the pattern for all the chemical treatments later used in making rubber.

In the summer of 1932, experimenters at the Westinghouse Electrical Company's laboratories in Pittsburgh, boiled two sausages by placing them into a powerful field of ultra-short radio waves. There were forecasts that this would lead to a revolution in the kitchen. Coal and gas would be exiled by radio cookers. But nothing domestic came of this radio-wave barbecue.

The microwave oven eventually arose out of war-time observations. In 1942 two British physicists, Sir John Randall and Doctor H.A. Boot, developed the magnetron, an electronic tube which produced microwave energy. This was soon put to use at radar installations and at one of these Percy Le Baron Spencer noticed the heat given off by one of these tubes. He tested the strength of this heat by putting a paper bag full of maize into the field of the tube. Within seconds the maize swelled, burst and he had perfect popcorn. So why not a tube-operated cooker? Spencer's firm, Raytheon, saw the commercial prospects and patented the first microwave oven in October 1945. But this was a large, heavy unit meant for big users, like hospitals and canteens. It was not until 1952 that the first household ovens reached the shops. These were made by the Japanese Tappan company. They were treated at first with great suspicion. No one at the time would have guessed that almost every other European home would have one within forty years.

The first lightning-conductor intended to protect a building was erected in September 1752 at the home of Benjamin Franklin in Market Street, Philadelphia. This followed a test of his theory, that lightning was electricity, by an admirer of his in France. This amateur scientist, M. d'Alibard, had read Franklin's views and set up an experiment to prove or disprove them. He erected an 80-foot-high iron rod at his country house and on 10 May 1752 this was struck by lightning. A Leyden jar was instantly charged up by the strike, proving that Franklin was right. After protecting his own house, Franklin arranged for the erection of lightning conductors on both the Academy Building and the State

House in Philadelphia. This marked their first use on public buildings. False teachings in Europe halted the use of these conductors there for seventeen years, until, in 1769 the first conductor was bolted to the steeple of St Jacob's Church in Hamburg. This was followed by the first use in Britain, when St Paul's Cathedral was fitted with a lightning-rod in 1770.

The first modern *Olympic Games* were held in Athens in 1896. The first day began on 6 April and the first Olympic Gold Medallist was James B. Connolly (USA) winner of the hop, step and jump contest. Only 311 competitors took part in this first of many Games, and of these 230 came from Greece! Over the years complaints were raised about the nationalistic fervours aroused by these Games and in 1936 a rival set of games was first staged in Barcelona under the title *The Workers' Olympiad*. The outbreak of the Spanish Civil War seems to have put an end to this rival organization.

Saccharine was first discovered jointly by Constantin Fahlberg and Professor Ira Remsen at the John Hopkins University in Baltimore in 1879. First to publish details of the discovery was Fahlberg, on 27 February 1879. The white crystalline substance they had created had no food value, but it was 550 times sweeter than sugar and a boon to slimmers and diabetics. It can be a danger to health, but only if consumed in huge quantities. For almost eighty-five years it had no competition as an artificial sweetener; then in 1965 it was joined by Aspartam, the new synthetic sweetener developed by the American Searle Laboratory. This is only 200 times sweeter than sugar, but it wins by having a more pleasant taste than any of the saccharines on offer.

The first mention of the famous *Angels of Mons* came in the afternoon and evening editions of the London *Evening News* for 29 September 1914. They appear in a short story called *The Bowmen*, written by Arthur Machen. Briefly, this story tells of a large-scale retreat of the British Expeditionary Force in Flanders. The fatigued troops are saved from destruction by the sudden intervention of St George and a long shining line of Agincourt archers. The ghostly archers cut down the advancing enemy troops and the day is saved. Ten thousand dead Germans lie scattered over the battlefield, but not one of them bears a single wound. Unmistakable fiction, one would have thought, but within days both the *Occult Review* and the equally occult magazine *Light* asked if the story was based on fact. Very soon the story started to circulate disguised as an 'eye witness account' and the phantom bowmen became changed into winged angels.

From July 1915, angels were in season with a vengeance. Sheaves of articles appeared in the papers. So did batches of heated correspondence. Many clergymen preached sermons using these angels as the central theme, and Machen sat back bewildered at this display of mass gullibility. Every time he tried to explain that his work was sheer fantasy, his voice was swamped by cries of 'It must be true'. But, although many 'witnesses' were talked about, not one came forward. Plenty of liars took the stage, but they were soon exposed. Machen wearily summed up thus: 'It has been claimed that "everybody" who fought from Mons to Ypres saw the apparitions. If that be so, it is odd that nobody has come forward to testify at first hand to the most amazing event of his life. Many men have been back from the front, we have many wounded in hospital, many soldiers have written home. And they have all combined, this great host, to keep silence about the most wonderful of occurrences, the aspiring assurance, the surest omen of victory.'

The first cremation in Britain took place on 25 September 1769 at the St George's Burial Ground, Hanover Square, London. The body of Honoretta Pratt was burned in her open grave at her express orders. She believed that ' . . . the vapours arising from graves in the church yards of populous cities must prove harmful to the inhabitants . . .' No action was taken to stop the cremation since the dead lady came from an influential family; she was the daughter of Sir John Brooks and the widow of the Honorable John Pratt, Treasurer of Ireland. The burning, though, had not been legalized by a court ruling.

The first *legal* cremation took place on a Welsh hilltop. On that site in 1884, the eccentric Archdruid Doctor Price of Llantrisant, placed the body of his son Iesu Grist (Jesus Christ in Welsh) upon a large oil-soaked, brush-wood pyre and began the cremation. But a crowd of chapel-goers rushed up the hill, seized Doctor Price; and only police action saved the doctor from a lynching. He was put on trial and the half-burned body of his son was used as evidence against him. But Price conducted his own defence arguing that cremation was healthier than burial. He won the case and the complete cremation of his son's body was ruled as legal.

On his death in January 1893 he left full instructions for the cremation of his own body. On the morning of 31 January 1893 the first officially held outdoor cremation took place at Llantrisant. Over twenty thousand people gathered for the ceremony, and for weeks afterwards, thousands of visitors were drawn to the small town. They even paid to visit the site of the cremation. Doctor Price had not lacked fame while alive, but his death gave him more fame then ever before. He is remembered with tenderness by generations of undertakers.

This free-lance stand by Price simply served to make people in Britain think again about cremation, but on the Continent of Europe, thoughts had already crystallized into action much earlier. The first logically designed cremation incinerator was actually brought into use in Padua on 10 March 1869, when a woman's body was cremated. The furnace used was designed by Doctor L. Brunetti. Later, his improved furnaces were admired by Queen Victoria's Surgeon, Sir Henry Thompson and he began advocating their use in Britain. He was instrumental in creating the Cremation Society in April 1874, but no practical steps were taken to introduce the method until 1879, when a crematorium was built at Woking, Surrey. But its furnaces stayed cold until 1885, due to opposition to the venture from the Home Office. The Price decision halted this opposition and on 26 March 1885 the first body, that of an unnamed woman, was reduced to ashes. The first crematorium in Europe had been functioning for nine years before this at Milan, where it was sponsored by the Milan Cremation Society. In the USA the first crematorium opened in 1876 at Washington, Pennsylvania. This was a privately owned establishment set up by Julius le Moyne, and open to all able to afford the fees. Public crematoria came much later.

The first of the Nocturnes, those romantic piano pieces, was created not by Chopin as many imagine, but by an Irishman working and living in Russia! The Irishman was John Field son of a Dublin violinist. As a child prodigy he had been taken under the wing of Muzio Clementi, the man held to be the first genuine composer for the piano. He travelled to St Petersburg with Clementi, and decided to stay there when his master left. Field was taken up by the aristocracy and in the 1800s he devised the Nocturne form, a slow piano piece with graceful embellished melodies for the right hand and broken chords for he left. It was the epitome of the romantic mood that suffused the aristocracy at the time. These examples were studied and admired by the young Chopin, who modelled his own nineteen nocturnes very closely on those of Field's.

The first motor cycle was a Daimler product of 1885 powered by a four-stroke, single cylinder engine, and first driven by Paul Daimler on 10 November 1885. This first trip was one of six miles, and was more of a testing spree for the engine than a public demonstration of a valuable new product. All further machines until 1894 were one-off, home-made devices; then the first commercially produced model appeared. It was the *Motorrad* made by H. & W. Hildebrand and A. Wolf of Munich. It used a 2.5 hp, 760 cc. engine with water-cooling, and this gave the bike a top speed of 24 mph. The first fifty production models were ready by

November 1894; and by 1896 over a thousand of these machines were on the road.

The first Boy Scouts originated at the Siege of Mafeking in 1899. In charge of the besieged township was British Army officer Robert Baden-Powell. The seven-months long siege led to a great deal of improvisation by Baden-Powell. A home-made gun was put together, to defend the place; an emergency issue of banknotes was designed and printed; and a set of special postage stamps was created and used internally. Young lads were given the job of ferrying the mail around on bicycles, and in between deliveries were trained to act as lookouts around the perimeter of the town. In other words they were trained in some of the basic arts of military scouting. Baden-Powell published his first work on scouting in 1899, this was *Aids to Scouting*.

The organized Boy Scout movement arose out of the memories of Mafeking. Its first activities took place at a camp on Brownsea Island, near Poole, Dorset. At this place, begining on 29 July 1907, Baden-Powell organized a troop of twenty boys into four patrols. They were instructed in observation techniques, night patrolling, knot-tying, fire-making, and a number of other army-derived skills. Over this camp flew the very flag flown over Mafeking during the long siege. This first camp was followed by the first book aimed at would-be scouts. This was issued in fortnightly parts under the title *Scouting for Boys*. Its author was, naturally, Baden-Powell and after the first part appeared on 16 January 1908, the seal was set on a new type of youth organization. Which of the troops was first to form is still in dispute, but we do know, that three troops were organized in Nottingham in the first week of February 1908.

By the end of 1908 the movement was formally organized on a national basis and the first permanent headquarters were opened at 116–118 Victoria Street, London. The instantly recognizable uniform hat was based on the army hats worn at Mafeking. The badge was also an army borrowing, based on the north compass point badge used by the 5th Dragoons. The motto '*Be Prepared*' actually meant '*be prepared to die for your country*'. This is made quite explicit in the first Scout writings, but deleted from later editions.

Despite this militaristic bias, the movement quickly drew young school girls into its net. At first these grouped themselves as Girl Scouts and the first unit was the Glasgow Cuckoo Patrol, founded summer 1908. Official approval was first given in 1909 with the publication of *A Scheme for Girl Guides*. This was printed in the November issue of the *Scout Headquarters Gazette*. But this approval meant the end of the Girl Scouts as an integrated part of scouting. They had to give up boisterous activities and develop ' . . . the ability to be better mothers and Guides to the next generation'. These stuffy words of Baden-Powell set the parameters for the new Girl Guides, who were placed under the control of an all-ladies committee ruled over by stern sister Agnes Baden-Powell. This first committee took charge at the beginning of 1910 and many guides then learned, for the first time, how to make tea. All very British.

The first game of table-tennis was invented by sportsman and engineer James Gibb, in 1889. At first it was simply a game played at his own home with corks instead of balls. When it came to marketing the

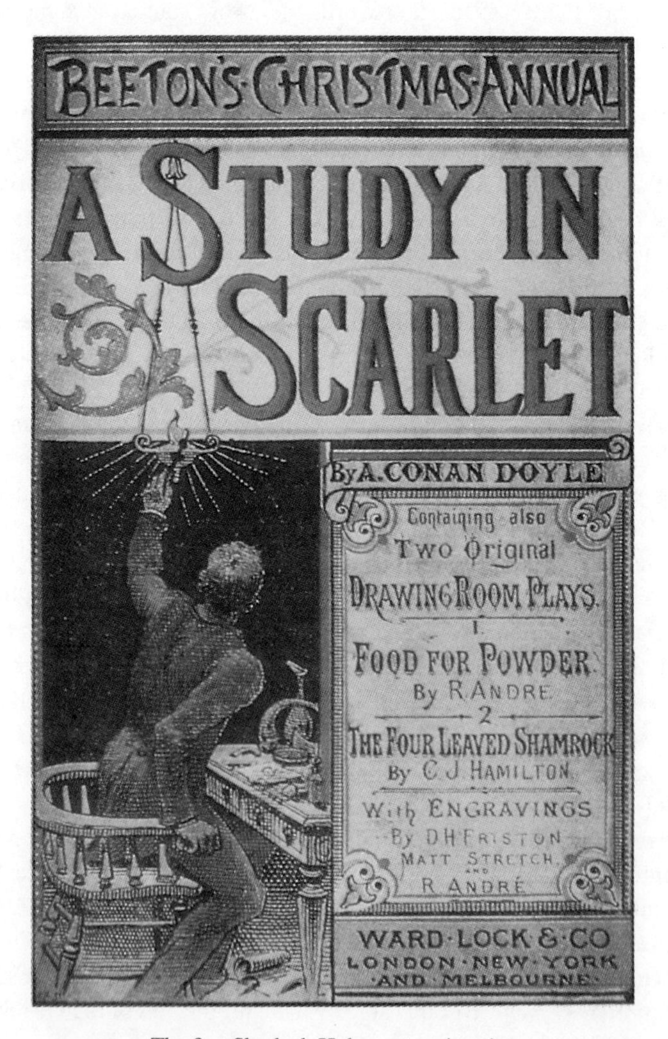

The first Sherlock Holmes story in print

game Gibb had hollow celluloid balls made and the games manufacturers John Jaques and Son saw its possibilities and began to supply it to Hamley's of Regent Street, London in 1898. It was first named *Gossima*, but this was dropped in favour of *Ping-Pong* and the game became a sweeping success, not just in Britain, but in the United States as well. The French hated it and thought it effete. But the rest of the world thought otherwise. The first Table-Tennis club was formed in the City of London in 1901 and met in Moorgate Street. In 1901 the first open tournament was held on 14 December. This was the Championship of London event held at the Royal Aquarium. The men's event was won by R. Ayling; the women's by Miss V. Eames.

The first work featuring Sherlock Holmes and Doctor Watson was 'A Study in Scarlet', first published in *Beeton's Christmas Annual* for 1887. Doctor Arthur Conan Doyle's notes for the 'A Study'... show that the first name given to his champion detective was Sherringford Holmes, while the first name given to Holmes' faithful companion was Ormond Sacker! Fortunately for us all, Doyle decided to rethink his plans and the now unforgettable names and characters appeared in that first story, and from then on. The dreadful name of Ormond Sacker was resurrected and first used openly in a Sherlock Holmes adventure in 1975. This was the Gene Wilder film *The Adventure of Sherlock Holmes' Smarter Brother*, where it was bestowed on a Watson-style assistant loyal to Holmes' younger brother.

Baked beans were an American favourite long before they were married to the tin and tomato sauce. The first recipe for the *original* form gives 1lb of pork to 1 quart of beans. The sauce, as such, was created by the pork juices and thickened with molasses. This recipe is from Boston in 1829, which explains the constant references to Boston baked beans in later American cook books. But the cult of the baked bean extended well beyond snooty Boston. Throughout New England farm folk made bean-baking a joyous ritual for weekends. Each home soaked the beans overnight, added the saltpork, crude molasses, mustard and herbs next day and sent their pots to the local bakehouse for a day-long simmering.

The first canned beans made by Burnham and Morrill (US) in 1875 stuck with the same type of seasoned molasses sauce. Tomato sauce was first used by the Indianapolis firm The Van Camp Packing Company, in 1891. The baked bean was first introduced to Britain in 1905 by the American firm of H.J. Heinz and became such a winner that Britons now eat double the amount (per head) as the American originators. Latest news on the bean front is that the age-old search for a wind-free bean is now at an end. For the first time new breeds have been developed which are trouble-free and promise to take us into a new golden age of bean gluttony.

The first public concert was organized by violinist John Banister on 30 December 1672. His daily concerts were reported on by Roger North, musician and Attorney-General, in these words: 'He procured a large room in Whitefryars (London) neer the Temple back gate, and made a large raised box for the musitians, whose modesty required curtaines. The room was rounded with seats and small tables, alehous fashion. One shilling was the price and call for what you pleased; there was very good musick, for Banister found means to procure the best hands in towne, and some voices to come and performe there, and there wanted no variety of humour, for Banister himself (inter alia) did wonders upon a flageolet to a thro' Base, and the severall masters had their solos.' These first concerts lasted for six years and were the first authentic *public* concerts in the world. Prior to this time all concerts were private affairs held in the homes or palaces of the rich.

The first sewing machine to be patented was the device planned by cabinet-maker Thomas Saint (17 July 1790). There is no evidence that it was ever made, but his specification shows a number of features found in later machines. It should be noted though, that he only thought in terms of a chain-stitch; which has great limitations, since if one stitch breaks, the whole line will unravel. The first working sewing machine was made by a French tailor, Barthelemy Thimonnier in the late 1820s. He patented his machine in 1830 and found no shortage of customers, but he never went beyond a chain-stitch apparatus. Despite this a battery of eighty of his machines was put to work on making uniforms for the French Army. He fell from grace when a Luddite fervour swept through a mob of Parisian hand tailors. They saw the labour-saving machine as a menace to their trade and they stormed Thimonnier's shop, smashed all his machines and threatened his very life. Meanwhile, in the United States, inventor Walter Hunt had devised the first US sewing machine; this had the defect of only being able to stitch short seams, just a few inches long. But it did, for the first time, make two important steps forward. It created a sequence of *lock stitches*, which would not unravel when the thread parted at one place. And it used a needle with an eye formed at the point. Hunt's machine, first made in 1832, was never marketed, for the inventor was one of those socially alert people who came to fear that mechanization would damage the lives of the handworkers. He withdrew his machine and never even patented it.

The next man to take up the quest for a versatile sewing machine was Elias Howe, a poor mechanic, completely lacking in Hunt's brilliance. He also knew nothing of Hunt's work, so started from scratch and toiled endlessly over problems long solved. His single-mindedness paid off when he found a backer willing to feed him and his family and provide a working space and tools. By May 1843 Howe had completed his first sewing

Singer's first sewing machine

machine. In July he gave the first demonstration of its work when he sewed the seams on two woollen suits. He then arranged the first public contest between machine and seamstresses at the Quincy Hall Clothing Manufacturing Company in Boston. He took on five seamstresses and completed his five test seams a little before them, and to a superior standard. His first patent was granted on 10 September 1846, but at this high point his backer lost interest and Howe floundered around trying to find someone to take up the rights.

For a while it looked as if all was well when an English manufacturer paid $1,500 for the English rights. The buyer, William Thomas, was a manufacturer of corsets and leather goods, and as part of the deal he insisted that Howe came to Britain and worked on a special model able to tackle leather-work. The first sewing machines made in England resulted from this alliance, which began in February 1847. But Thomas turned out to be a devious character. He treated Howe not as a partner but as an ordinary workman; and he conveniently forgot the fact that he had promised Howe royalties on all sales.

Howe broke with Thomas and soon found himself destitute. He borrowed money to return home and began a bitter fight against the many manufacturers who were now infringing his patents. One of those taken to court was Isaac Singer who had invented his own special machine and patented it on 12 August 1851. It was more like the machine we know today than Howe's, and should rightly be acknowledged as the first machine using modern principles. But the needle used had an eye at the tip and the courts decided that this one feature alone was proof of infringement. Singer was ordered to pay $25,000 royalties to Howe. Then Singer, in his wisdom, went to Howe and proposed that all known sewing machine patents should be pooled, with Howe drawing royalties from all future machine sales. Howe agreed and the first Patent Pool came into being. Singer forged ahead, became a household name and the first man to spend a million dollars a year on advertising.

The first safety razor was devised by American King Gillette and patented on 2 December 1901 (No. 28,763 1902 UK). King had first thought of the idea in 1895. A cutting edge backed up by just a minimum of supporting steel would work as well as a cut-throat razor but would need no stropping or regrinding, since it could be thrown away when blunt. The first hurdle lay in making such a blade. Steel manufacturers scoffed at the idea, but Gillette's mechanic, William Nickerson, proved them wrong. The American Safety Razor Company began producing their first razors and blades for the USA in 1903, and their first for Britain in 1905. From a ludicrously slow start the shaving habits of the whole world then became changed for good.

Life without paper would be full of inconveniences but this material of great utility is a comparative newcomer, since it was first invented in 105 AD. The Chinaman who invented it is unnamed, but a certain Ts'ai Lun is the first man known to have reported the invention to the reigning emperor Ho Ti. Microscopic examination of the earliest known Chinese paper shows it to be made of rag fibres, the material used for centuries for the finest papers (and still used). Without paper, printing would never have flourished. Vellum and parchment were all very well for limited editions but quite unsuited for the *News of the World* or the *Washington Post.* Though we know that the Chinese and Koreans printed from wooden blocks, we see that their ingenuity stopped one stage too soon. Their blocks were of single words, and as such were just a variant on an illustrated block. Printing from moveable letter types is acknowledged to have been first carried out by Johann Gutenberg in Strasbourg in the 1430s.

The first dated piece of printing is a papal indulgence of 1454, while the first dated book is a Psalter of 1457 which carries the first printers' names as Johann Fust and Peter Gernsheim. In 1469 appeared the first printed book-list from a publishers. This was Fust and Schoeffer's single sheet list of twenty-two of their offerings. The first printed book in the English language was actually printed on the Continent in about 1474. Its printer was William Caxton, a Kentishman who served the English King as 'Governor of the Merchant Adventurers at Bruges'. Its title was *Recuyell of the Hystoryes of Troye.* Place of printing was Bruges. The first dated book published in England was Caxton's *The Dictes or Sayengis of the Philosophers* of 1477. The first to feature woodcut illustrations was the *Myrrour of the World*, another Caxton production of 1481.

Apart from being the first printer, Caxton must count as well as the first English editor. Until the late eighteenth century all printing involved type, woodcuts, etchings and engravings, then in 1798 came the first new printing process first called 'polyautography', then renamed lithography. This invaluable invention was due to the experiments made by Aloysius Senefelder of Munich. He was seeking a cheap method of home printing from etched plates, and to cut costs this involved grinding and repolishing the plates after all its wanted impressions had been taken. This was impractical, so he tried etching the surface of flat slabs of Kellheimer stones. On one of these stones he wrote some lists using a wax-based black ink. Over and over again he tried new ways of picking up the inked writing and at one stage he noticed that if ' . . . there happened to be a few drops of oil in the water into which I dipped paper inscribed with my greasy stone-ink, the oil would distribute itself evenly over all parts of the writing, whereas the rest of the paper would take no oil'.

The essential principle had been found. A limestone slab written on by a waxy crayon, then moistened with water, would only pick up a greasy ink where the wax lay. First patented in 1800, this new system was quickly adopted for posters and illustrations, in either black and white, or full colour. Photographic images were then used to create cheap, exact copies of old masters, popular paintings and advertising copy. In 1875 the first offset lithographic printing method was used by Robert Barclay. He picked up the inked image from the stone slab and transferred it to a smooth surface wrapped around a revolving cylinder. This cylinder, in turn, transferred its image to a metal sheet which then received the paper. The offset method became widely used to print designs on thousands of different tin containers, on tea trays, on metal advertising signs and on tin-plate toys. Practically all posters today are printed by the offset method, so are many of the world's postage stamps. And forgers too, find it very alluring.

The first safety pin was an invention of American Walter Hunt, the same fellow who made the first sewing machine to make lock-stitches. He patented his *Dress-Pin*, as he termed it, on 10 April 1849 (No. 6,281 US). A thankful world salutes him.

Chocolate was first used as a drink only and originated in the West Indies in the seventeenth century. Its first use in Britain was first announced in the *Public Advertiser* of 16 June 1657. It was then sold, either as a drink or as paste for home brewing, at 'a Frenchman's house' found in Queen's Head Alley off Bishopsgate Street. The basic paste was made by crushing the roasted cacao seeds with arrowroot, or sago, and sugar. This gave a tasty but rather greasy drink, since the beans were rich in cocoa butter.

The first factory to produce *chocolate bars* was opened at Vevey in Switzerland in 1819. The creator of the bars was François-Louis Cailler. Cailler's first bars were of plain chocolate only; then in 1875 Cailler's son-in-law, Daniel Peter manufactured the first milk chocolate bars. In Britain the first promotion of chocolate in all forms was in the hands of the Quaker families of Cadbury and Fry. Being opposed to alchohol and stimulants they had naturally favoured something as inoffensive as chocolate as a drink and then as a sustaining food. The first eating chocolate was promoted by Fry's as ' . . . a pleasant and nutritious substitute for food in travelling . . .'. This applied to their Chocolate Lozenges first introduced in 1826. John Cadbury's first offering was their 'French

Eating Chocolate' of 1842. This was the first solid chocolate to be sold as a pleasurable thing to eat and not as an emergency ration. Cadbury's were also the first to introduce chocolates in boxes in Britain in 1866. Their first boxed assortment comprised; almond, orange, lemon and raspberry soft centres. In the same year Cadbury's became the first to introduce cocoa to Britain. Their Cocoa Essence of 1866 was a pure powder with all the objectionable cocoa butter removed. And it may well have been the first pure cocoa ever made, since the first Dutch cocoa of 1828 (van Houten's) still had a fatty residue left after milling.

The first instant coffee was made in the form of essences created by passing steam through ground coffee made from well-roasted beans. This was reduced to a thick syrup, often with the addition of sugar, honey or glycerine, and bottled. This product first made in the late 1800s, could also be turned into a powder form by evaporating the essence, preferably in a vacuum pan. This was never carried out as a commercial success until 1937; the first attempt, in the US in 1867, was a failure. The 1937 success was the result of some eight years of research by the Nestlé firm of Switzerland. Their Nescafé started an irreversible trend in people's drinking habits, though it should be noted that their first 'instant coffee' was a misnomer, since for years it was mixed with lactose. Thus the first real instant *coffee* did not emerge until the 1960s.

Postal envelopes and wrappers, officially prepaid for national use, were first placed on sale on the same day as the first adhesive stamp, the Penny Black. This was 6 May 1840, and then the public could buy their first Mulready envelopes or wrappers in two prices: one penny black or twopenny blue. The Mulready design showed Britannia dispatching winged messengers to all parts of the world. For some reason this seemed a ridiculous theme to many and the Mulready alternative to the adhesive stamp was soon discarded without much mourning.

The British Ordnance Survey was first created by a ledger entry made on 21 June 1791. This Expense Ledger of the Board of Ordnance recorded the payment of £373.14s to Jesse Ramsden for a 3-foot theodolite. At that time there was an urgent need for military maps that were large-scale and reliable. This need had been identified as early as 1763 by William Roy. Britain was then at loggerheads with France and Roy saw how vulnerable the country was to invasion. Any worthwhile plan for defence could only be made if the lie of the land at any given point was known with great accuracy. With that in mind, cartographer Roy made his first proposal for a 'general Survey of the whole Island at public cost'.

The Paris Observatory

It was turned down as a ' . . . Work of much time and labour, and attended with great Expence to the Government.' By 1784 relations with France had improved to the point where cross-channel co-operation between the observatories of Greenwich and Paris was established on a very friendly basis. At one point this led to a query from the French about the exact latitude of Greenwich. Such a question could only be settled by precise surveying and George III supplied the funds for the work, the Royal Society organized it, and Roy found himself in charge. His experience with this work came to the notice of the third Duke of Richmond, a man with strong cartographical interests – he even had his own band of surveyors on his estate. In 1782 Richmond became Master General of the Board of Ordnance and on Roy's death in 1790 took over his cause and became the actual first founder of the Survey. In 1800 the military importance of the Survey was recognized by the establishment of the first army cartographical unit, the Corps of Royal Military Draftsmen, based at the Tower of London.

The first banknotes to be signed with a *nickname* were the notes issued by the National Bank of Cuba after Fidel Castro came to power in 1959. The much venerated Ernesto Guevara was Castro's second-in-command at the time, and became president of the National

Bank. A new banknote series was designed showing scenes from the revolutionary war and Guevara dispensed with custom by signing them with the nickname everyone knew – 'Che'.

The first clarinet was invented by the great Nuremberg woodwind maker J.C. Denner in the first years of the eighteenth century. Prior to its invention there had been a small instrument called the chalumeau which used a single reed mouthpiece like the clarinet, and had a cylindrical bore, also like the clarinet, this is first mentioned in 1687. But this chalumeau was not capable of being used for anything but simple music. The clarinet, by contrast, has a great range since its 'speaker key' allows the player to overblow the fundamental notes a twelfth higher. Extra keys fill the gaps between the two registers and the instrument then offers the greatest range of any of the woodwinds, with each register having its own special quality. The clarinet is first mentioned in a Nuremberg document of 1710. The form we now know best, the Boehm system, was first devised in Paris and patented in 1844.

The first stethoscope was invented by a Breton doctor in 1816. It arose out of embarrassment on the part of Doctor Rene Laennec, since he found it blush-making to have to press his ear against the bosoms of his female patients. At the time this was standard practice when chest conditions had to be diagnosed. So he began by using a rolled up paper tube placed between ear and bosom and found that he heard much better that way. The first treatise on the *Stethoscope* (Greek for 'I look into the chest') was published in 1819 in French and 1825 in English. Those first used were made as monaural stethoscopes, single tubes turned out of boxwood or ebony. Then ivory tips were fitted and decorations were added. The binaural type was first devised by G.P. Caniman of New York in 1850 and first patented by him in 1855. In this form it became the shape we all recognize at once, though the monaural form is still used in maternity wards. Apart from doctors others found good uses for the stethoscope. It was first used in 1888 to listen to the first wax cylinder recordings, both at home and in amusement arcades. And used-car salesmen have been using them since the 1950s.

The kaleidoscope was first invented by Sir David Brewster in 1817. Originally he employed two trapezoidal mirrors placed in contact at their wide ends, to form a dihedral angle. The object to be viewed was placed at the larger end and viewed from some 6 to 10 inches away. The viewer then saw graceful symmetrical forms resulting from the multiple reflections. Later versions used sealed tubes with an assortment of coloured objects (paper, glass, feathers) contained inside. These could be shaken up to form endlessly varied sets of patterns. Still later versions used up to four mirrors arranged as a hollow prism and the effects became even more dazzling. Kaleidoscopes still fascinate, even in an era of electronic games. While in the cinema its principles have been employed to create special effects many times.

The first bikini was displayed to the public on 3 June 1946 as part of the swimwear collection of Louis Réard, of Paris. He named his ultra-abbreviated coverings *Bikinis* after the Bikini Island in the Pacific. His slick explanation for this choice was that his coverings were as revolutionary and explosive as the Atomic bomb detonated on the island a few days earlier. The first person seen in public in the bikini was the dancer Micheline Bernardi. Her normal place of work was the Paris Casino, not a catwalk, but none of the professional models would display themselves in Réard's 'straps', so she was called in to take over the job. She probably wondered what all the fuss was about, since for most of her working hours she was covered in nothing but pink lights and glory.

The Pill, the influential female contraceptive, was the result of a commission from the Planned Parenthood Movement in the USA. They wanted to see a 'harmless, entirely reliable, simple, practical, universally applicable and aesthetically satisfactory' type of contraceptive. This was a tall order, nevertheless, Doctor Gregory Pincus of the Worcester Foundation for Experimental Biology took up the search in 1950. Five years work on, an oral contraceptive led to a pill based on progestin and oestrogen. First clinical tests took place in 1954, then in 1956 came the first large-scale proving tests. In San Juan, Puerto Rico a group of 1,308 women volunteers took part in these tests. Two different forms of the Pill were used and after three years and only seventeen pregnancies, the scientists were satisfied that their goal had been reached. Even so the Pill was not launched without further refinements; then on 18 August 1960 *Enovid 10* became the first oral contraceptive to be commercially made. This was a product of the G.D. Searle Company of Illinois. In Britain the first available Pill was *Conovid*, launched on 1 January 1961.

The first parking meters came into service in Oklahoma City on 16 July 1935. They were the bright idea of journalist Carlton Magee who served as the part-time chairman of the Business Traffic Committee set up in 1933. Many ideas for controlling parking were considered, then Magee suggested the meter system. It was accepted and a Dual Parking Meter Company was set up. Its dual aim was to control traffic congestion and raise revenue at the same time. In Great Britain the need for such meters was not felt until the late 50s, then on 10 July 1958 the Westminster City Council brought the first British meters into operation and the age of nonchalant parking came to an end.

Who made the first barbed wire is still in dispute; but we are sure that barbed wire as we now know it, was first patented by Joseph Glidden in 1874 (No. 157,124, US). Before that date we find barbed fence wires by Lucien Smith, and William Doninson Hunt both patented in 1867; and William Kelly's diamond barbed wire of 1868. Originally meant to fence in the huge cattle ranches of the Wild West, it later became feared as the hateful barrier strewn across battlefields. Its first use in warfare came during the Spanish-American War of 1898, when the US Army used it to protect its lines and stores.

The first crossword puzzle was invented in 1913 by Arthur Wynne of the *New York World*. That newspaper had been keen on games ever since 1890 when it printed the *Nellie Bly* board game on its front page; 'Cut this out . . . paste it on cardboard and play according to the simple directions'. This first newspaper board game set the tone and Wynne's idea was taken up in the issue for 21 December 1913. He based the crossword on the memory of an old Victorian parlour game 'Double Acrostics'. But his game placed black squares between the word spaces and gave thirty-two clues to the player. It caught on fast, so much so that in 1924 the first crossword puzzle book was published by Simon and Schuster Inc., of New York. In this original form, the crossword was a

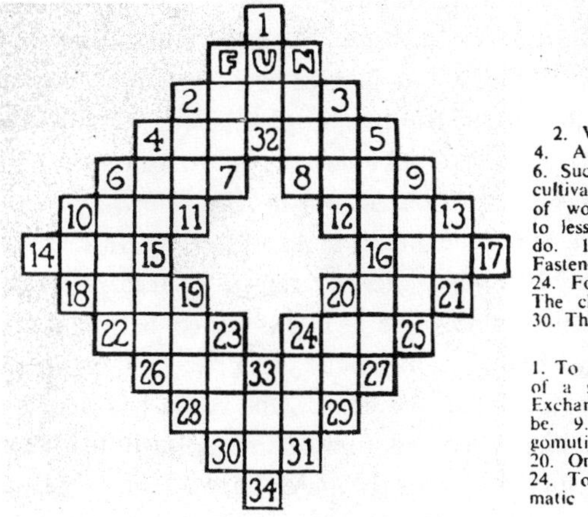

CLUES—ACROSS

2. What bargain hunters enjoy. 4. A written acknowledgement. 6. Such and nothing more. 8. To cultivate. 10. A bird. 12. A bar of wood or iron. 14. Opposed to less. 16. What artists learn to do. 18. What this puzzle is. 20. Fastened. 22. An animal of prey. 24. Found on the seashore. 26. The close of a day. 28. Elude. 30. The plural of is.

DOWN

1. To govern. 2. A talon. 3. Part of a ship. 4. A daydream. 5. Exchanging. 6. What we all should be. 9. Sunk. 10. The fibre of the gomuti. 13. A boy. 19. A pigeon. 20. One. 23. A river in Russia. 24. To agree with. 33. An aromatic plant.

The world's first crossword puzzle

quite simple puzzle, it only became cryptic after it was first introduced to Britain in November 1924, by editor C. W. Shepherd. Shepherd's first appeared in the *Sunday Express* on 2 November 1924 and within a year the complex British style of puzzle was being devilishly concocted by *Torquemada*, a tantalizing fiend who aptly chose the name of the most feared of all the Spanish Inquisitors.

Aspirin, or acetylsalicylic acid, was first prepared by Doctor Karl Gerhardt in 1853. But a form pure enough for pharmaceutical use was not synthesized until 1897. This was the work of Doctor Felix Hoffman of the German Bayer Company. He heated salicylic acid with acetic anhydride and patented the process in 1898 (No. 27,088). In Germany the product was first sold in May 1899 in powder form under the trade name *Aspirin*. In Britain Bayer's Aspirin was first imported in 1905.

Dynamite was first manufactured by Alfred Nobel in 1867. His factory at Helenborg, near Stockholm was then engaged in manufacturing nitroglycerine. The nitro was unmatched for its explosive power, and was in demand for blasting operations. But it was so sensitive that handling it raised great problems. These problems were solved when Nobel used 25 per cent of the siliceous earth called Kieselguhr to absorb the nitroglycerine. The resulting thick sludge could be cast into stick form and handled without fear of premature explosion. First called Nobel's Safety Powder it was soon given the trade name *Dynamite.*

Millions of American baseball fans have no doubts about the origin of their game. For them it was first invented by army officer Abner Doubleday in 1839. At West Point, the story runs, he devised a game based on a diamond-shaped pitch, with a base at each corner, which the batsman had to run to, in due order. But this 'original' game was already in existence in Britain long before Doubleday hit his first homer. The game of *Rounders*, first described in detail in *The Boy's Own Book* of 1828, when compared with Abner's game, is almost identical. In fact the few differences are too slight to make the two games separate and distinct. They are one and the same. The real difference between the American game and its British original lies in the sheer pace and immense skills called for on American pitches. In the US the game has reached peak development. The game there was first taken seriously by 1845, when the first organized clubs began to compete. In 1876 the National League was founded and in 1901 the American League was set up to rival the National. From this rivalry grew the World Series games first held in 1905. In this series the champions of each league meet annually to compete in seven games. The rules they play under are the Cartwright Rules, first used in New Jersey on 19 June 1846. And there is a strong feeling among many fans and players that the maker of those rules, Alexander Cartwright, is the true father of American Baseball.

The first *Meccano* set was invented by Frank Hornby of Liverpool in 1900. It was first meant to amuse and instruct his two sons. By using metal strips with holes punched at regular intervals he gave his children a great choice of home-made structures. When bolted together the strips could make cranes, houses, bridges, carts; the list was endless. Extra pieces like brass pulleys and hooks made the constructions even more exciting. And working models were possible as well. When he decided to market it, he used the stodgy name *Mechanics Made Easy* but in 1907 this was dropped and the winning name *Meccano* was first used. His invention still lives on, though Hornby himself is best remembered for the clockwork train sets he later produced.

Radar (Radio Detection and Ranging) was first thought of in the 1900s and the 1904 patent of Doctor Christian Hulsmeyer for, '. . . employing a continuous radio wave to detect objects' is the first small step towards the goal. It is now claimed that John Logie Baird's patent of 21 June 1928 (No. 292,185) covers Radar by describing '. . . a method of viewing an object, by projecting upon it electromagnetic waves of short wavelength adjacent to the infra-red radiation in the spectrum, but of longer wavelength than the infrared rays, exploring the object or the image thereof by a device sensitive to such rays . . . and traversing a spot of light across in synchronism with the exploration of the subject.' This patent was actually applied for first in 1926 and there is little doubt that Baird himself viewed it as the primary radar outline.

But Radar as it came to be used, owes its conception to Sir Robert Watson-Watts who first began work on the radio location of aircraft in 1935. This came about after he had been asked if a 'Death Ray' was possible. The enquiry came from the Committee for the Scientific Survey of Air Defence, who were soon told that the idea was science-fiction. On the positive side, reported Watson, electromagnetic waves might be used to locate enemy aircraft and set them up for destruction. He gave the first demonstration of the infant Radar system on 26 February 1935 when he traced the flight path of a Heyford bomber from eight miles away. For this experiment he was able to call on the short-wave transmissions of the BBC's giant aerial at Daventry. A portable cathode-ray oscilloscope registered the traces caused by the aircraft.

This first improvised test led to the first purpose-built establishment in

June 1935. At this base near Orford, in Suffolk, they extended the tracking distance, first to 17 miles, then to 40 miles. The defensive value was proven and in December 1935 the first permanent Radar stations were ordered to cover the Thames Estuary. First of these to be operational was the Bawdsey Manor station which opened in May 1937. Then a further chain of stations was ordered along the East Coast and these stations (twenty in all) were placed on complete 24-hour watch on Good Friday 1939. It has been said that the Battle of Britain might well have been lost if Radar had not been there to forewarn. It should be noted that the term Radar was not used in Britain until 1943; prior to that date it was officially *RDF* or *Radio Direction Finding*. The name Radar started out as an American usage, first employed by Commander S. M. Tucker of the US Navy in 1940.

Margarine was first invented as an entry for a prize competition initiated by Napoleon III. The Emperor was seeking a substitute for butter that would be suitable for the poorest strata and for use by his Navy and Army. Basically he was after a buttery-tasting fat that had good keeping powers. The one and only entry for this prize came from Hippolyte Mège-Mouriès, who spent over two years in experimenting and ended with a mixture of suet, skimmed milk, and pork and beef offals. With some seasoning and much mixing, this gave an agreeable fat, pleasant enough in taste and smell, but dead white. A little yellow colouring took care of this wan look later on, but in its first, pearly state, the product took the prize and was patented on 15 July 1869 (No. 2,157).

The inventor called it *Margarine* because of its whiteness, basing his trade-name on the Greek word for pearl, *margaron*. The first factory dedicated to manufacturing Margarine was opened, not in France, the Franco-Prussian War put paid to that, but at Oss in Holland. This venture was the work of Anton and Jan Jurgens who bought the manufacturing rights from Mège-Mouriès. In Britain the new product was first sold as *Butterine*, until pressure from farming interests made the name illegal. In the United States farming pressures made the very colouring of Margarine illegal in a number of States. Distributors got around this restriction by offering the first colour-it-yourself packs. The white fat was packed in a tough transparent pack, along with a dense yellow hued blob of paste. The buyer could knead the pack in the hand until this colour was uniformly mixed in, and there it was, a pack of imitation butter, just as if the factory had supplied it.

The first large full-scale solar furnace was constructed at Meudon, in France in 1946. The plant was able to reach temperatures of 3,000 degrees Centigrade but this example never sparked off a worldwide interest. Research switched instead to collecting solar energy through the medium-ship of cells working on two main principles: the thermo-chemical, and the photovoltaic. The photovoltaic solar cell uses the same technology used in powering hundreds of the satellites in orbit. The first cells were developed in the Bell Laboratories by US scientists D. M. Chaplin, C. S. Fuller and G. L. Pearson. When first developed in 1954 these cells had a relatively low rate of efficiency, just 6 per cent coupled with a high cost. By 1978 efficiency had risen to the point where these single crystal silicon cells could convert 17 per cent of the sunlight hitting them into electricity. And multicrystalline cells, though less efficient (12 per cent) can be packed more tightly into an array and increase their value just through this one feature. Cheaper, though down in efficiency (5 per cent) are the amorphous photovoltaic cells, made by depositing an

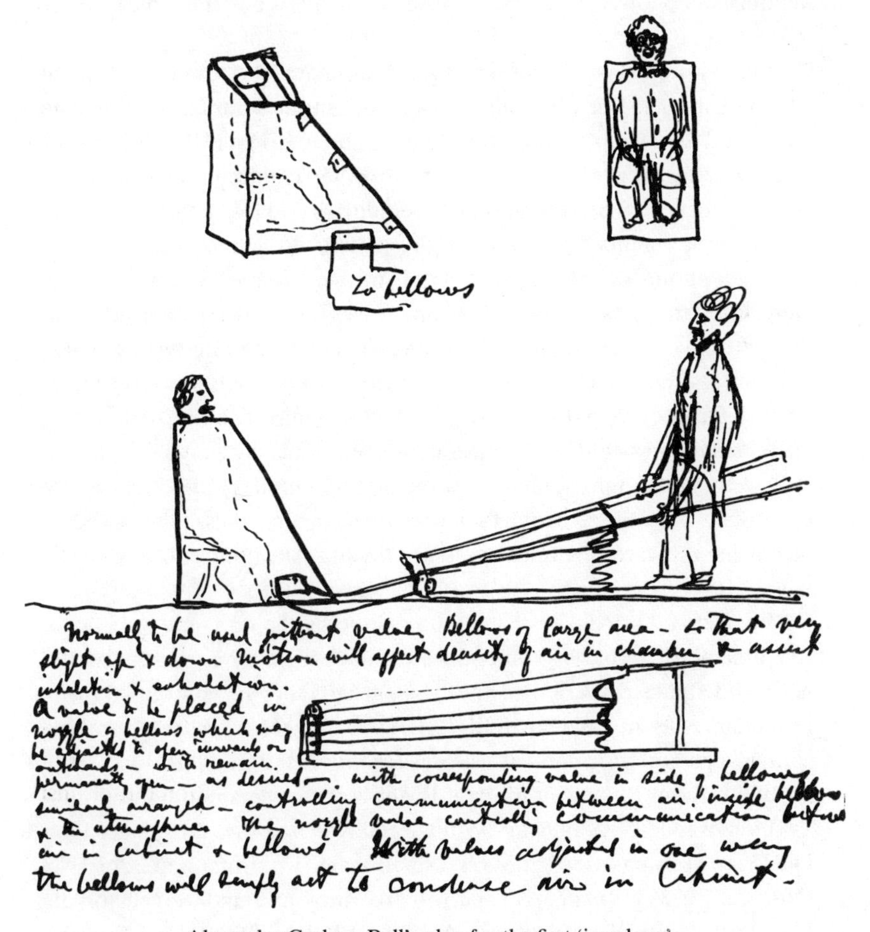

Alexander Graham Bell's plan for the first 'iron lung'

ultra-thin film of molten silicon on to a thin sheet of stainless steel or glass. This type of cell was first introduced in 1983. By 1987 it was so perfected by Sanyo of Japan, using their new Amorton technology, that the efficiency yield was raised to very high levels. Since this was accompanied by a drop in production costs the photovoltaic cell may well become usable by many who now manage without any electricity whatsoever. Some 60 per cent of the world's population have no access to power grids and yet there is this great asset, sunlight, with its enormous energy going to waste. The only rival to the photovoltaic cell is the Thermochemical cell, developed by the Coal Research Division of the Max Planck Institute in West Germany. First made in 1988 by the Bominsolar Company, this cell can transform up to 80 per cent of the solar energy falling on it into electricity. Since its costs per unit of electricity are said to be much lower than its rivals, the future would seem to belong to it. But its cells are a much bulkier than photo-cells and that, at the moment, is a real handicap.

The first step towards the *Iron Lung* was taken by Alexander Graham Bell, after the death of his newborn son in the summer of 1881. The infant had died from a respiratory failure, and in his grief Bell looked for some way to prevent this happening to other children. He invented a 'vacuum jacket' to cope with such emergencies and his sketches show an airtight iron chamber surrounding the patient up to the neck. A hand-operated pump was meant to be fitted to the chamber and when operated would then rhythmically raise and lower the air pressure inside, and this in turn would compress and expand the lungs. Despite this advanced thinking, nothing seems to have been done to provide such apparatus until 12 October 1928. On that day Professor Philip Drinker (of Harvard, USA) tested his iron lung at Boston Hospital. This first modern artificial lung was a great improvement over the first trial model of 1927, since that had to be improvised from two vacuum cleaners; these alternated in producing the carefully timed pressure fluctuations on the thorax.

The *Aerosol* principle was first discovered by the first man who shook up a bottle of bubbly mineral water, stuck his thumb over the bottle-neck and squirted away. But the first patent using carbon dioxide as a propellant was not issued until 1903 (No. 746,866 US). A disposable spraycan suitable for aerosol use was first invented on 22 August 1939, by Julian Kahn of New York. But this invention remained unused until 1941 when Lyle Goodhue and William Sullivan first saw spray cans as an ideal way of using insecticides on a small scale. Their first patent of 1943 (No. 2,321,023, US) sums up the research they had undertaken for the US Department of Agriculture. At first only benefits seemed to come from their use and the diversity of aerosol-packed products became astonishing. Then the doubts crept in. Today their role in damaging the ozone layer has been accepted and the system is in rapid decline.

The first magazine for women was *The Ladies' Mercury* published by John Dunton of London. Starting in June 1693, this double-sided sheet specialized in problems of the heart and social manners. It was, in fact, one long agony column of questions and answers. The first all-ladies magazine, that is actually edited and planned by a woman, was *The Female Tatler*, first seen in July 1709. This was edited by Mrs Mary de la Riviere Manley, a lady with strong views on the scandalous goings-on among the rich. Some of her revelations were undoubtedly invented to please her readers, a technique still much in evidence today. Unashamed fiction first arrived in *The Records of Love,* a short-lived magazine of 1710. Fashion, as you may guess, was first catered for by the Parisian magazine *Le Cabinet des Modes* of 1785. It takes little imagination to guess the intended readership for the *Maids', Wives' and Widows' Magazine* of 1832. This was the first to try to cover a wide range of topics: fashions; domestic economy; poetry; reviews of plays and literature and other wholesome offerings. But despite this fare the magazine was never troubled by mass-circulation. First to count as a mass-circulation periodical was *The Englishwoman's Domestic Magazine*, first out in May 1852. This was edited by Samuel Beeton (now best known as Mrs Beeton's husband) and was meant to improve the intellect and morals of his readers. So it is odd to realize that it is best remembered for its lengthy correspondences on such topics as the tight-lacing of corsets and flagellation!

The first electric iron was the invention of a supreme optimist. He was Henry Seeley of New York who patented his iron on 6 July 1882. It was, however, little more than a gift to the future, since houses were then not wired up to electric power supplies. In fact the first electric power station did not come into action until 3 p.m. on Monday, 4 September 1882. This was the Edison Electric Illuminating Company's plant at Pearl Street, New York. It had a grand total of eighty-five customers.

The first microscope was invented by a Dutch spectacle-maker in 1590. Zacharias Jansen used a concave lens as his eye-piece, in conjunction with a convex lens as the objective. This is the simplest form of the apparatus and was soon replaced by the first compound microscope, said to be the work of Cornelius Drebbel.

This is reputed to have been displayed in London in 1621. It differed from the simple style since it used a *field lens* placed between the objective and eye-piece. This improved light-gathering and definition.

The first book written about discoveries made with the compound form, was Robert Hooke's *Micrographia* of 1665. The first great discoveries made with the aid of the microscope were those of Anthony van Leeuwenhoek. He ground such fine high-powered lenses that he was able to see micro-organisms for the first time. At last men became aware of a sub-miniature world that could possibly have some bearing on their aches, pains, fevers and exits.

We have to thank the San Francisco Gold Rush of 1850 for the wagon-train of events that led to the first *Jeans*. Bavarian immigrant Levi Strauss headed for the territory along with huge bales of stout cloth, meant for tents and wagon covers. But others were there before him with the same ideas, and so he was landed with a mound of unwanted fabric. Then the bird of inspiration flew. Levi heard a miner complain that his trousers were quickly torn and rendered inelegant by the harsh, rocky diggings. Before the man could rush off to Savile Row, Levi was there, with the toughest pair of pants ever seen. His tenting material was the answer to a miner's prayer, even if it meant walking a little stiffly. And plenty of miners agreed and flocked to his improvised tailors' tent. His fortune was made. The familiar rivets were an afterthought, first introduced in the 1860s as a pocket reinforcement. The blue hue was first used simply because that was the colour of the batch bought by Levi.

The first Nobel prizes were awarded in 1901, but the first Nobel medals were not distributed until 1902, since they were not minted in time for the first-ever ceremony. The medals are of 23-carat gold, 2.5 inches in diameter and weigh nearly 8ozs. On one side is a bas-relief portrait of Alfred Nobel, on the other side are classical figures and a Latin quote saying 'How good it is that the life of man should be enriched by the arts he has invented'. Each winner receives a unique, individually designed diploma of great beauty. The first, and only, exception to this rule was made in the case of the 1915 Physics Prize winners. The Prize went jointly to the father and son team of Sir William Henry Bragg and Sir William Lawrence Bragg, and this special occasion was marked by the first issue of two identical diplomas. Only the illuminated names differed.

The first inventor of roller skates lived to regret it. He was Joseph Merlin, a Belgian maker of musical instruments who decided to show off his invention at a grand masquerade held at Carlisle House, Soho Square. This affair was meant to be the high spot of the 1760 season and Merlin made his grand entrance into the ballroom with a panache that stunned everyone. Not content with just gliding in on wheels, he further added to the spectacle by playing a violin solo at the same time. It was all too much. With both hands occupied he lost all sense of control and careered into the huge ballroom mirror. It smashed; the violin smashed, and the skater sprawled battered and bleeding in the debris. There were a few more brave attempts after this, but the roller skates, as we know them, with their four wheels, were first patented by James Plimpton of New York in 1863. Roller skating quickly became a craze that animated Americans and Europeans for decades to come. And public skating rinks were erected to cater for the fans. But violins stayed out of favour.

Spectacles are first mentioned at the end of the thirteenth century, and the first spectacle maker is believed to be Salvino Armati of Florence. The date of the invention can be worked out from a passage written in 1305 by Fra Giodano di Rivalto who says, 'It is not yet twenty years since the art of making spectacles, one of the most useful arts on earth, was discovered'. Those first spectacles were made with convex lenses and useful only for long-sighted people. The first known use of concave lenses for short-sight can be dated at 1571, since at that time Raphael painted Pope Leo X, and the Pope's spectacles show the concave curvatures without doubt.

For centuries spectacles were made to clamp on the nose; then in 1728, the London opticians, Edward Scarlett, made the first spectacles fitted with stiff side-pieces. These were the so-called *Temple-Spectacles*, presumably because they appealed to the army of London lawyers. In 1752 the first hinged side-pieces were introduced and the designs began to look more familiar to our eyes. At that time, lenses could be clear or tinted to relieve the eyes in bright weather. And there were even special sun glasses, though these went back to 1579 when lenses of amber, soaked in linseed oil to make them transparent, were first use to protect the eyes from glare. The versatile bi-focal was first invented by none other than Benjamin Franklin, who grew grouchy whenever he had to swap lenses when working. So he placed lenses of the two strengths he needed, together in one frame, and managed by looking through the right lenses at the right time.

Bi-focals, with their compound lenses made in one piece, were first developed and made by the Carl Zeiss company in 1910. The attempt to dispense with spectacles altogether, began with Sir John Herschel in 1845; but he was way ahead of the technology of the time. Once again it was Carl Zeiss who made the first breakthrough when they used the research of Doctor A. E. Frick to design and make their original contact lenses in 1877. Zeiss used optical glass for their contacts, then in 1936 the first *Plexiglas* contacts came from the laboratories of I. G. Farben. Until 1945 all such lenses covered the whole eye, then from the USA came the first tiny lenses designed to sit on the cornea. Nine years later saw the introduction of the first flexible hydrophilic lenses; a development by Czech scientist, Doctor Wichterle. This led inevitably to the present-day use of short-life disposable lenses.

In the search for Firsts one of the fascinating things is the regular appearance of some long-forgotten claimants, just when the matter seems to be settled. Out they come, from their caves, their tar-paper huts and other places of retreat. They all wave proofs of their priority. One such hermit is Nathan B. Stubblefield. He claimed to be the first to demonstrate wireless. Not just morse signals, but intelligible speech and recognizable music. And all this a full three years before Marconi sent out his first crude signals in Italy. In 1892, Nathan, a Kentucky farmer, demonstrated his apparatus in Philadelphia and New York. He then took it to Washington DC and made the first ship-to-shore radio broadcast from a boat anchored off Georgetown on the Potomac River. The world's first *Wireless Telephone Company* was then set up and a golden future seemed certain. But the inventor was secretive; no one was permitted to examine his apparatus; no patent was ever applied for, and the company was wound up. His apparatus then mysteriously disappeared and Stubblefield became a recluse, in the best American tradition. He died of starvation in 1928 in his remote hillside cabin in Kentucky. Had he really been first? Without knowing how his equipment worked it is hard to say. He may well have been employing an induction system and not a true wireless system. In which case he was simply travelling over well trodden ground, since the basic principles were long established; and it was well known that an induction system would only work over short distances. But since Stubblefield is still hailed as the first (mainly in Kentucky), his work has to be acknowledged and perhaps re-examined. Incidentally, the induction system still has its uses in conference halls, museums and other places, where people need to pick up information without being hampered by trailing wires. A wire loop around the rooms induces speech into headset receivers with remarkable efficiency.

The first agreed standards of electrical measurements emerged from the International Electrical Congress held in Chicago in 1893. They agreed that the Ohm (unit of resistance) should be defined as the resistance offered to a steady current by a column of mercury measuring 106.3 cm long, with a mass of 14.4521 grams, with a uniform cross section and a temperature of zero degrees centigrade. The *Ampere* (unit of current) was a littler simpler. That was defined as the unvarying current which would deposit metallic silver from a solution of silver nitrate at the rate of 0.001118 grams per second. This, of course, is very exciting to know. Even more exciting is the realization that this eminent body defined the *Volt* wrongly. Their first standard would not fit in with the requirements for the Ampere and Ohm, and caused confusion. This was only resolved by the Electrical Congress of 1908 which decided on the first logical definition of the Volt. (*Watts* and *Joules* and *Henrys* and *Farads* are great fun too.)

The first man to reach the North Pole is held to be Commander Robert E. Peary of the United States Navy. The date of his arrival at the Pole is recorded as 6 April 1909. And on 7 September 1909 the *New York Times* published the first news of his exploit received in a cable which read: 'I HAVE THE POLE APRIL 6. EXPECT ARRIVE CHATEAU BAY SEPTEMBER 7. SECURE CONTROL WIRE FOR ME THERE AND ARRANGE EXPEDITED TRANSMISSION BIG STORY.' The

Doctor Frederick Cook, the first to reach the North Pole?

big story promised at that time did not arrive until 10 September, but by then some rather curious news had been carried by the New York *Herald*. On 2 September the *Herald* published a dispatch from 'the first man to reach the North Pole', but the arrival date quoted there was 21 April 1908, and the claim was made by one of Peary's former colleagues, Doctor Frederick A. Cook.

For years afterwards a bitter war of words was declared between the Peary camp and the Cook camp. A string of highly detailed, argumentative books kept the battle raging. Books with provocative titles like: *The Great North Pole Fraud*, *Did Peary Reach The North Pole?*, *Has The North Pole Been Discovered?*, *Peary Did Not Reach The North Pole*, *Peary At The North Pole: Fact Or Fiction?* and *The Case For Doctor Cook*. At the time the controversy first began Peary's case was favoured by the American establishment. Peary had many rich, influential friends and this was used against him by those persuaded by Cook's writings and lectures. But, despite the arguments, Peary was given the Grand Gold Medal of the National Geographic Society 'for the discovery of the North Pole'. A $10,000 purse was given to him by the Peary Arctic Club, and the Navy raised him to the rank of rear admiral with an annual pension of $6,000.

In recent years Wally Herbert, leader of the 1986–89 British expedition that achieved the first surface crossing of the Arctic Ocean, came to conclude that Peary had been from 30 to 60 miles away from the Pole. A later study, carried out by the National Geographic Society, places Peary much nearer to his goal. This new study makes extensive use of photo-analyses of pictures taken at the time. Despite this, the champions of Doctor Cook still hail their man as first. Yet, even if the claims for Peary are correct they are still wrongly phrased, for Peary was not alone at the North Pole. With him was Matthew A. Henson. So why no joint-acclaim for Henson and Peary? And there you have a problem, for Henson was a black American, and in the America of 1908 it was unthinkable to place white and black on the plane of equality. In fact it took seventy-nine years and a petition to the President of the United States before Henson's remains were placed alongside Peary's in the Arlington National Cemetery. On 6 April 1988 the reinternment ceremony finally acknowledged that the conquest of the North Pole had been a shared triumph for both Matthew Henson and Robert Peary.

The terms vaccination and inoculation are now used as if they are synonymous, but at one time there was a clear distinction. Inoculation once implied the introduction of a very small amount of small-pox virus into the bloodstream of a fit person. A slight bout of the disease would follow and bring about future immunity to a severe, perhaps fatal, bout. Vaccination, by contrast, sought to create immunity by a much safer method. Matter taken from the hands of milkmaids infected with cow-pox was placed into scratches made on the arms of the healthy. The immunity to the related small-pox was thus gained without the pain and danger of passing through a real, even if minor, attack of the disease. This preventive measure was the result of observations showing that milkmaids who had contracted cow-pox never succumbed to dreaded small-pox.

The vaccination technique was first used medically by Doctor Edward Jenner who knew of the milkmaids and their strange protection, and by applying his reason worked out a systematic way of treating patients. His first complete experiment was carried out in May 1796, when he took 'Matter from the hand of Sarah Nelmes who had been infected by her master's cows, and inserted by two superficial incisions into the arms of James Phipps, a healthy boy of about eight years old. He went through the disease apparently in a regular and satisfactory manner'. So far so good. Then came the real test. Would the boy go under if infected with matter from a small-pox pustule? The boy's parents must have had supreme faith in Doctor Jenner, for they allowed him to place small-pox matter into incisions made on their child's body. This first crucial test was made on 1 July 1796 and the boy stayed healthy, unaffected by the deadly germs. The case was proven. In 1798 Sampson Low of London, published the first account of Jenner's discoveries; an account written by Jenner himself.

You may never use or even see a chronometer for the whole of your life, and yet the chronometer has played a little-noticed part in all our lives. Without this invention there would have been no accurate methods of navigation. This was recognized by all the maritime powers in the sixteenth century and prizes were offered to the first man able to provide a watch that would keep exact time on long sea voyages. At sea the ship's navigator could mark the instant that the sun came to his meridian and call it 12 o'clock. The difference between this, and Greenwich time on an exact watch would then give the Longitude. A similar calculation could be reached by using a given star.

The need for a watch that would stay constant, even with temperature changes, grew so great that in 1714 the British Parliament set up its first committee to consider this question. An Act of Parliament was then drafted promising huge prizes in gold for the first watch to meet their specifications. The reward could be as high as £20,000. Among those who entered for this contest was John Harrison, a carpenter whose first working watches were made in wood. In 1736 he completed his first naval chronometer and it was first tested on a man-of-war bound for Lisbon. The captain's report was favourable, though the great goal was still in the distance, and Harrison was given £500 to continue his work.

Three improved watches later, Harrison's son took ship for Jamaica, with Watch No. 3 slung in position between gimbals. Its 7-inch dial gave William Harrison supreme confidence. When the ship's crew swore that the cursed thing was leading them astray, Harrison replied that their maps were wrong, not his father's workmanship. And he was right. The Portland Islands were found by his reckoning, not by the inaccurate charts. The ship was able to take on fresh stores and Harrison, and watch became joint heroes.

On reaching Port Royal the watch was found to be five seconds slow after eighty-one days. The reward was claimed. A further gruelling voyage proved his case yet again; and the sum of £20,000 was awarded in two parts. It was then 1762; the work had taken him all of forty years to reach success.

Even so, his watch was soon replaced by an even finer one made by professional watchmaker John Arnold. In 1772 he made his first marine timekeepers and with his later modifications these became the standard patterns that governed naval use for centuries. Arnold, who used jewelled movements, was also the first pioneer of the precision pocket watch. His watches are prized by collectors internationally.

Penicillin was first discovered in 1928 as a result of the accidental contamination of an experiment. The place was a laboratory at St Mary's Hospital, Paddington. The experimenter was Alexander Fleming who was simply growing bacteria on small glass dishes as part of his studies in staphylococci. Fleming's lab was a trifle on the grubby side, and from some grubby corner, or some grubby window, came spores of a mould which landed on one of his dishes, and grew. The greenish mould was eventually spotted by Fleming, who was most indignant, until he saw that the mould had killed off the bacteria. The mould was identified as Penicillium notatum, and Fleming concluded that some substance secreted by it had slain his cultures. He named this unknown substance *Penicillin*, but experimented little and gave up work on it when it proved difficult to isolate.

Ten years later Howard Florey, Ernst Chain, and their colleagues at Oxford, began work on Penicillin as a therapeutic agent. If it could kill off staphylococci then it was potentially invaluable in treating infected wounds and septic conditions. Florey and Chain found the best moulds; found ways of creaming off the Penicillin from the dross; and designed vessels in which the mould could be effectively grown. Then wartime needs led to the involvement of hundreds of scientists in their work. By 1944 enough penicillin was in production to treat all the serious casualties sustained during the invasion of Europe.

In 1945 Sir Alexander Fleming, Sir Ernst Boris Chain and Lord Howard Walter Florey were given the joint Nobel Prize for their work. To be fair, though, it was Florey who first showed the real value of Penicillin, and pursued its development and purification. By comparison Fleming seems like an idle observer of a mildly interesting piece of serendipity.

The first Impressionist Exhibition was opened at Nadar's Photographic Studio, at 35 Boulevard des Capucines, on 15 April 1874. 165 paintings were on show, including works by Renoir, Sisley, Degas, Monet, Cézanne and Pissarro. Among the twenty-nine artists on display was the first woman to embrace Impressionism, Mademoiselle Berthe Morisot, who showed nine of her works. Though now recognized as the first assembly of Impressionist art, the exhibition catalogue did not use the term. This apt and appealing name was first used on 25 April 1874 by Louis Leroy in a satirical article of his published in *Le Charivari.* Before then other journalists had written about the 'quality of impressions' at the exhibition and this turn of phrase was suggested by exhibit number 98; a small canvas by Claude Monet entitled *Impression, soleil levant* (*Impression, Sunrise*). But it was Leroy who first spoke of a school of 'Impressionists' and the artists came to accept the label, even if it was first used in a derisory manner. Yet, for a while, the new movement had an alternative label; they were called 'Intransigents'. Ernest Chesneau even wrote as if this was the title that would stick to them: '. . . this school has been baptised in a very curious fashion with the name of the group of Intransigents'. But this name had strong political overtones; it was derived from the name given to a group of Spanish anarchists. Given the memories of the recent Paris Commune, with its bloody episodes, such a name was most unwelcome. So, in a way, the artists were glad to adopt a term that was politically neutral and, by their third exhibition in April 1877, they were content to speak of themselves as The Impressionists. But not one of their catalogues, for all eight of their exhibitions, ever used the term in any context whatsoever.

One of the curses of the past was the prevalence of a fatal infection following childbirth. Puerperal sepsis, as it was termed, was first shown to be preventable in a treatise written by Charles White in 1773. He was an obstetrician practising in Manchester and his observations led him to advocate fresh air, cleanliness and antiseptic measures in all cases of childbirth. His pamphlet *The Management of Pregnant and Lying-in Women* failed to change attitudes. Those were the days when surgeons would come straight from post-mortem rooms to attend patients.

In the United States the first man to campaign for cleanliness in the fight against puerperal fever was Oliver Wendell Holmes. But when he argued his case before the Boston Society of Medical Improvement, he was met with indifference and hostility. Professor Meigs, the most famous name in midwifery in America, even claimed that this childbed scourge was an act of God. Fifty years later Holmes wrote: 'Others had cried out against the terrible evil before I did, but I think I shrieked my warning louder and longer than any of them – before that little army of microbes was marched up to support my position.'

Independent of Holmes, one of the most tragic figures in the history of surgery took his stand against the barbaric standards found in hospitals. In Vienna Ignaz Semmelweiss stood appalled by the strange standards accepted in the great Maternity Hospital. In the wing used only by midwives the fever rate was low; only one per cent died. But in the wing used to train medical students the rate was ten times higher. Post-mortem examinations to determine the nature of the fever led them nowhere. In fact the rate went up after each exploration finished. And no one thought of washing hands, or changing clothes. It was only when Doctor Kolletcha died that Semmelweiss realized just how their research work was killing the women. Kolletcha had been nicked by a student's post-mortem knife. The knife was smeared with matter from a woman killed by the fever. And the autopsy on the professor showed the same symptoms as found in puerperal sepsis. Nothing was known of bacteriology then, in 1847, but Semmelweiss knew that putrid particles were being carried by the doctors to the unfortunate women. All students were then compelled to wash their hands in chlorinated water before examining the pregnant women. Absolute cleanliness became the rule, and he enforced this doctrine with fanatical intensity.

Even though the death rate dropped dramatically, other obstetricians refused to accept his reasoning. He went on to prove that the infection could be passed between living patients by contact and even spread by soiled bed-clothes. But the jeering and sneering still went on and Semmelweiss lacked the stamina to take this for long. His mind became affected. He wrote letters to the leading obstetricians, denouncing them as murderers. Finally, with his mind in turmoil, he was admitted to an asylum where he died of septicaemia, brought on by a scalpel cut, the same type of accident that had led him to his great discovery in the first place. Semmelweiss, first man to eliminate this hellish scourge, first man to bring aseptic conditions into hospital work, died on 20 July 1865.

In the same year, in England, Joseph Lister began his campaign for antiseptic conditions in surgery. He did not discover antiseptics as such, but he was the first to discover the principles behind their use in preventing and treating sepsis in wounds. His system aimed at eliminating the invading germs by attacking them with carbolic. Hands had to be washed in carbolic, instruments steeped in the fluid, even the dressings had to be soaked in it. And after all that was done, there was one extra care to be taken; the very air in the operating room had to be sprayed continuously, with a fine spray of a carbolic solution. Lister's first successful case involved a patient with a compound fracture. He saved the limb.

The first account of his work was published in *The Lancet* in 1867 and first details were given by him on 9 August 1867. On that occasion his audience was packed with the delegates to the thirty-fifth annual meeting of the British Medical Association. It seemed to be the ideal place for mature deliberations, but as soon as Lister finished his address he was attacked by eminent surgeons, among them Sir James Young Simpson, first man to discover the properties of chloroform. Lister was accused of stealing the researches of French apothecary Jules François Lemaire, and his techniques were said to have been tried by others and discarded by them as useless. An examination of Lemaire's book of 1863 shows that the Frenchman had never developed a method of treating wounds but, at the time, Lister's attackers were taken seriously. His famous, influential opponent Simpson drove Lister into a position of isolation. But Lister persisted with his campaign and in 1868 developed the first use of absorbable catgut ligatures for tying off blood vessels.

Even this major step failed to win him recognition in Britain, but in Germany his ideas were applied and treasured. In 1867 they were first acted on in Leipzig by Doctor Karl Thiersch. Gangrene and pyaemia were eliminated at his clinic. The Berlin Charity Hospital then followed Leipzig's example. Leading surgeon Richard von Volkmann of Halle and Professor von Nussbaum of Munich became converted to Listerism and banished the spectre of putrefaction from their hospitals. And it was in Germany that a small-town doctor came to prove, for the first time, that living microbes bred the gangrene and fevers that Lister had battled against. His name was Robert Koch. Lister had fought against invisible enemies; Koch now made them viewable under the microscope and even photographed them. Now it was possible to wage war in different ways, on different foes, for the first time.

The first non-stick pans were put on sale by the French Tefal Company in 1955. But the coating they used, *Teflon*, had been developed years earlier by Doctor Roy Plunkett of Du Pont of the USA. He discovered Teflon, or to be exact polytetraflourethylene (PTE), in a casual fashion while working on refrigerants in 1938. Apart from being a good insulator, this PTE was inherently very slippery and that suggested a whole host of possible commercial uses. But no one at Du Pont thought of using it on cooking utensils. That bright idea came to Marc Grégoire in 1954. He first used some PTE to lubricate his fishing tackle and then saw that the stuff could be spread over the base of a frying pan to provide a non-stick surface. His Tefal Company grew out of this brainwave and is still the leading company in this field.

Linoleum, the first mass-produced floor covering, was invented by Frederick Walton and patented by him in 1863 (No. 1,037 and 3,210). Others before him had tried to make a covering using cork fibres, canvas and oil paint, but Walton's *Lino* was the only factory-made product that caught on. In making it he used cork dust and resin, mixed in with oxidized linseed oil. This goo was spread on to a backing of cotton or flax and allowed to dry out. It became the first choice for those who wanted a cheap, hard-wearing floor cover. A room could be covered from wall to wall at a fraction of the cost of the cheapest carpet. And a great choice of designs was available, since the Lino was easily printed with coloured patterns, or pictorial motifs. Until the advent of vinyl floor coverings, Lino was without serious competition and dominated for almost a century.

The potato crisp, which seems to spawn a new flavour every month, was first created as an act of defiance. In 1853 at Saratoga Springs, New York, a customer at Moon Lake House Hotel complained that the fried potatoes were far too thick. The short-order cook was an American Indian called, unromantically, George Crum. But if he had lost his tribal name he still preserved his fighting spirit and he hit back with a batch of fried potatoes made from the thinnest slices anyone had ever seen. The crisp chips were crunched with contented grunts; shouts of 'encore!' showed that a new dish had been born, and the rest is history, as far as the United States is concerned. In Britain, though, crisps were latecomers. The first were only seen in London in 1913 when Carter's Crisps were put on sale. Manufacturer Carter had sampled crisps in Paris and saw their potential. The little blue paper twist of salt found in crisp packets for decades (now re-introduced) was first thought up by Frank Smith, who started his famous firm in 1920. The twist was there to prevent people walking off with the salt cellars in the shops where the crisps were first sold.

In 1944 DNA, or deoxyribonucleic acid, was first shown to be the substance that transmits hereditary characteristics from one bacterium to another. This was demonstrated at the Rockefeller Institute in New York, by scientists Oswald Theodore Avery, Colin MacLeod and Maclyn McCarthy. In 1953, James Dewey Watson and Francis H. Crick first described the exact structure of DNA in a joint article in *Nature*. They showed that the DNA molecule took the form of a double-stranded helix formed by a series of nucleotides. This work earned them the 1962 Nobel Prize for Physiology and Medicine.

Since it was shown that the DNA patterns for each individual were unique, it soon became evident that this knowledge could be used in forensic medicine. At Leicester University Professor Alec Jeffreys worked out a system of genetic fingerprinting and this made legal history when it was first used to convict a criminal in November 1987. The charge at that time was rape and, at Bristol Crown Court, Robert Melias was found guilty and imprisoned, after it was shown that his DNA was identical with that found in the sperm samples left on his victim. Following that came the first murder conviction in which DNA evidence was offered. Killer of two teenagers, Colin Pitchfork was sentenced to life imprisonment in 1988. And in July 1988 DNA evidence was admitted for the first time into a American murder case, when Timothy Wilson Spencer, the 'South Side Strangler', was charged with the killing of Susan Tucker. Spencer was convicted of this murder and three others. In many ways DNA promises to be an anti-crime aid even greater than that of fingerprinting.

The first lady tennis player to deliberately display her ankles was Lottie Dodd. In 1887, when winning her first Wimbledon title, she was aged fifteen, and that allowed her to get away with wearing a schoolgirl's calf-length dress. An older female displaying so much leg would have been regarded with horror. Lottie had one more advantage over her older opponents; she could dispense with a corset, and this gave her an unmatchable flexibility and speed. Beyond doubt her well-ventilated ankles and unfettered spine helped her win her five titles.

The first Viewcam video-camera was introduced by Sharp of Japan in March 1993 and launched in the UK in September 1993. The Viewcam differs from all previous camcorders because it discards the traditional viewfinder and uses a 4-inch miniature television screen to view the action. This is rotatable and fixed to the back of the squat camera. It provides a full-colour picture registered by a liquid-crystal display using 115,100 pixels, or tiny colour dots, each one of which is divided into four parts. All this gives a high resolution of detail. The screen can be used for immediate playback after filming, and it can even be adapted to work as an ordinary television receiver or as a video recorder.

The first Viewcam

Using the Viewcam gives film makers a new freedom. The machine no longer has to be held up to the face, and can be used comfortably at waist level. And a twist on the screen allows the operator to put himself in the action and control the camera remotely. The top-price version even comes with a special tripod that houses a voice-activated pointing device. This breaks with other remote control systems and frees all involved from being tied to the camera in any kind of way.

The first *Osteopath* was Missouri doctor Andrew Taylor Still, who began practising his system of manual manipulations, particularly of the spine, in June 1874. He believed that many aches and pains were caused by mechanical problems of the osteomuscular system. A systematic treatment, involving stretching, massage, even pummelling, would rectify these problems. His ideas were then codified and formed the curriculum of the first school of osteopathy, the American School of Osteopathy of 1892. A rival system of manipulation named *Chiropractic* first came into being at the hands of manipulator Daniel David Palmer in 1897. Palmer founded the first school of chiropractic in Iowa and insisted on a more vigorous method of manipulation than that advocated by Still. Both systems have their staunch advocates and both systems have their staunch critics. Few critics dismiss the systems in their entirety, but they maintain that the theories behind both methods are unsound. When manipulations work they do so despite the theories. Qualified physiotherapists, on the other hand, can deal with the same type of problems, in a much calmer fashion and without employing a jargon which is in dispute.

Ever-useful cellophane was first created by Swiss scientist Jacques Edwin Brandenberger in 1911. It was the result of a ten-year-long search for a waterproof wrapping material for perishable foods. His triumph came after he began work on viscose, a material first used in work on artificial silk (Rayon) in 1891. The original research had concentrated on making threads from the material, but Brandenberger aimed at making thin, flat sheets. His first usable sheets were ready in 1909, but these were just laboratory samples. It took two more years before he had reached the stage of certainty, and his first patent was issued in 1912 (No. 3,929). Even so, there was still work to be done before a completely waterproof form was achieved and this work was taken over by Du Pont in the USA, who patented cellophane in its finalized, waterproof form, in 1928 (No. 283,109).

In recent years we have heard a great deal of the merits and demerits of enzyme-enriched washing compounds. But the idea of using enzymes in this fashion is quite an old one. As Proctor and Gamble state in their patent declaration of 25 April 1967: 'Enzymes have been used as cleaning aids for many years. As early as 1915 Rohm found that fabrics could be cleaned more easily and at lower temperatures when pretreated with fat and protein digesting enzymes ... Later, in 1932, enzymes were utilized in a soap composition having greatly improved cleansing action ... Enzymes aid in laundering by attacking soil and stains found on soiled fabrics. Soils and stains are decomposed or altered in such an attack so as to render them more removable during laundering ...' Proctor and Gamble's patent is the first to specify the use of enzymes in washing powders. The 1915 patent referred to (No. 283,923) covers Doctor Rohm's first discovery of the usefulness of enzymes in laundering; while the 1932 patent by George Frelinghuysen (No. 1,882,279) applies to the first use of enzymes in a soap composition. Proctor and Gamble were later to be the first to introduce a machine-wash liquid. This was their *Vizir* of 1982.

The first carpet sweeper was patented by James Hume in 1811 (No. 3,399). He used a revolving brush placed in a box on castors. Ledges inside the box caught the dust as the bristles revolved. But this device had little impact. The first successful carpet cleaner was then invented by a china-shop proprietor of Grand Rapids, Michigan. Melville R. Bissell was unfortunately allergic to the straw he had to use to pack his china. It flaked easily and threw out dust, and ordinary brooms simply propelled this dust into the atmosphere of the packing room. Bissell designed a wheeled, lightweight, rectangular body which housed a spring-loaded cylindrical brush. A long handle allowed it to be pushed around with ease, while a container inside the body collected the dust picked up by the rotating brush. This was patented by him in September 1876, and then marketed by the first carpet sweeper firm, the Bissell Carpet Sweeper Company. Very soon the sweepers themselves became known as *Bissells*.

The golf ball as we know it, with its innards made from rubber threads, was first patented by Alfred Julius Boult in 1898 (No. 17,554). And in 1900 the Haskell Golf Ball Company, founded by Coburn Haskell of Cleveland, Ohio, patented the first machine able to automatically wind the rubber thread around a central core. It should be noted that the basic idea of such a ball was first hit upon by Haskell. Before his intervention golf was played with a solid gutta-percha ball, first introduced in 1848. Earlier than that, the balls had been constructed from hand-sewn leather cases stuffed with feathers.

The first milking machine made on sound principles was devised in 1895 by Doctor Alexander Shields. He had observed the problems created by the pioneer milking machine, made by fellow Scotsman William Murchland. Murchland was on the right track, but his patent of 1889 (No. 15,210) only provided for a *constant* suction at the cow's nipples. This was not only painful for the creatures but led to inflammation, which affected profitability. Doctor Shields saw that a pulsating control would give an intermittent suction and permit the nipple to contract and expand, just as it would when suckled by a calf. The first American milking machines of L. O. Colvin (1878) pre-date the Murchland machines, but were never commercially attractive.

The first hovercraft-inspired *Flymo* lawn-mower was produced in Britain in 1963. It was manufactured under the patents of the Swedish firm, Aktiebolaget Flymo (No. 929,610), whose 1963 specification covers: 'Blades shaped to give downdraft to support mower', and a blower driven by the motor which '. . . produces an air cushion below the housing to support the mower clear of the ground'.

The first *Christmas Card* was designed by John Calcott Horsely RA in 1843. But the first man to suggest the idea was Sir Henry Cole.

Apparently Cole needed a quick way of greeting his friends at Christmas time; his work schedule left no time for letters. The cards, sentimental and cosily family-centred, were lithographed by Jobbins and Co. of Warwick Court, Holborn, as black and white line drawings. Colour was added by hand by out-workers.

Both Cole and Horsely were rather prudish types so they were amazed when their cosy greetings cards were denounced as encouraging loose living and debauchery. This came about because the family group depicted was shown in the act of drinking a toast. In some people's minds wine and ale were inevitably linked with drunkenness and the discarding of wholesome family values. Such an outcry must have hit hard, for Cole was an ardent educational reformer, and Horsely was noted for his campaign against the use of nude artists' models. Indeed this campaign had even earned him the nickname 'Clothes-Horsely'. With time this stupid reaction was forgotten, but it still took some time before the Christmas card was accepted. In fact, it was not until 1868 that the first large-scale commercial production began. This was the work of Charles Goodall and Sons of London, whose cards, designed by John Leighton, made use of the now standard motifs of holly, snow and robins.

For ages British posters were crudely printed and designed by nameless artists of little merit. The first British poster to be designed by an artist of note was first seen on the billboards in 1871. It was a striking and haunting design showing a mysterious woman stepping out into a night lit only by stars. It was intended to advertise the dramatized version of Wilkie Collins' thriller, *The Woman In White*. The artist was Fred Walker and the wood engraver who cut the poster blocks was W. H. Hooper.

The Times, often regarded as the foremost newspaper in the world, first appeared under another name on 1 January 1785. It was initially burdened by the clumsy title *The Daily Universal Register*. Its proprietor, John Walter promised that his paper '. . . in its politics will be of no party'. He also showed that he was keen on raising revenue through advertising, writing: 'Due attention should be paid to the interests of trade, which are so greatly promoted by advertisements'. Oddly enough his venture was prompted by a belief that he could make his fortune through exploiting a new, patented method of typesetting, called *logography*. The newspaper was to be the living proof of the excellence of this new method. But it was the method that died while the newspaper, which was secondary to his plans, went on to glory. On 1 January 1788 it dropped the first title and became *The Times*, or to be strictly accurate,

The Times became its bold-type name while, below this, readers were reminded of the former name.

The first weapon to be called a sub-machine-gun was the *Thompson gun*, designed by an American General but world-renowned as the deadly 'Tommy Gun' toted by American gangsters. This was originally thought of as a trench-clearing weapon for use in the First World War, but the war ended before the design was finalized, and the gun first went on sale in 1920. With a fire-rate of up to 100-rounds-a-minute it was a formidable killing device at close quarters, and eminently portable. But it was not the first such gun to be used. As early as 1915 the Italians had introduced a light machine-gun using 9mm pistol ammunition. The drawback of this Italian design was the need to mount the gun on a tripod. The first sub-machine-gun to be free of such an encumbrance was then developed by the German Hugo Schmeisser as the MP 18.1, and first used by assault troops in March 1918.

The first man to make a living from breaking wind in public was a French entertainer Joseph Pujol. His stage title was 'Le Pétomane', and his act took him from provincial flea-pits to the Moulin Rouge itself. He discovered his strange talent at an early age and first turned it into entertainment in the 1880s. His début at the Moulin Rouge came in 1892 and was an overnight success. After all, how many people can imitate gunfire, smoke cigarettes, play tunes on a tin flute, and sound mock bugle calls, without once opening their mouths? It was said of him that 'Sarah Bernhardt drew box-office receipts of 8,000 Francs, but Le Pétomane in a single Sunday took 20,000 Francs at the box office'. Though the laughter was hysterical at times, Joseph Pujol brought a droll dignity to the conclusion of his act by farting the opening bars of the French National Anthem. What more could one ask for?

The first of the universally celebrated Gilbert and Sullivan Operas was staged at the Gaiety Theatre, London, on 26 December 1871. This was *Thespis; or The Gods Grown Old*, described as an entirely original Grotesque Opera in two acts. Action took place on Mount Olympus and involved a complete corps de ballet as well as the sixteen main players. It ran for sixty-four performances and led to the demand for more and more. In all seventeen comic operas were created by the talented team of Sir Arthur Sullivan and W. S. Gilbert (the S stood for Schwenck!).

The first cover for *Punch,* drawn by artist Richard Doyle, appeared on 6 January 1849. It was also his last! It proved so popular that it ran for the next 108 years and became the best known cover in the world. Its cheeky portrait of Mr Punch and dog together with their impish attendants, concealed some very naughty references, by Victorian standards that is. If the Queen had spotted them she would not have been amused.

Richard Doyle's first and last cover for *Punch*. It ran for 108 years

The first electric clock was patented in 1841 (No. 8,783) by Alexander Bain and John Barwise, ten years after the first electric motor had been demonstrated by Joseph Henry at the Albany Academy in the United States. Professor Wheatstone was busy trying to develop an electric clock at the same time as Bain, but the 1841 patent gives Bain priority of invention. His clock worked on direct current; the only form available at that date.

The first clock designed to work on an alternating current was the Warren Clock patented in 1919 (No. 125,766). In an attempt to break away from the limitations of mechanical movements, tuning forks were first used to control timekeepers in 1919. William Eccles then used a thermionic valve circuit to keep a tuning fork in non-stop vibration, thus giving a constant, regular pacing. In 1956 this principle was applied for the first time to a watch by the Bulova Company. Their patent (No. 761,609) shows a tiny tuning fork controller kept in motion by a transistor circuit. The first Bulova watch made on these principles was marketed as the *Acturon* and was guaranteed to keep time with an accuracy of one minute a month, when worn.

An alternative method of timekeeping had been worked on, as early as 1920. This involved the use of quartz crystal resonators to set the pace. In 1929 Warren Marrison (US) made the first efficient clock using a quartz resonator, but his work seemed to have little impact on the clock-world, which stayed with old, well-tried methods. The first fully electronic clock did not emerge until 1965 (No. 995,546) when the Swiss Vogel company brought out a clock without a single moving part; its time was displayed on a cathode ray tube. Watches, governed by quartz chips were first introduced by Seiko in 1969 and the quartz control principle quickly became the most important in international horology. Solar batteries came to be offered as an alternative to the replaceable silver batteries used to power the watches. Then, in 1988, came a development which dispensed with the battery altogether, when the first dynamo-operated watches were introduced by Seiko, Japan; and Jean d'Eve, Switzerland.

The first traveller's cheque was introduced by *American Express* on 5 August 1891. *American Express* were also the first to introduce the credit card to Britain on 10 September 1963. But they were not the first to invent the credit card. That was the work of Ralph Schneider's *Diners Club* of New York. Cards issued by *Diners Club* gave credit at twenty-seven restaurants and were first used in 1950. The idea was taken up by the first bank in 1958, when the Bank of America introduced its *Bankamerica Card*. In Britain the first bank to follow suit was Barclays with its *Barclaycard* of 1966.

The first parachute to use a ripcord was patented by the Irving Airchute Company in 1920 (No. 138,059). The specification states: '. . . upon entering the airship the aviator fixes the rope to some part thereof near where he sits when operating his machine. When he wants to jump out, he does so and the rope, by its attachment to the cover (of the chute), rips the cover off at the cords or threads which hold it in position.' This is the method still used by paratroops and formation jumpers. It is a misunderstanding to think that a jumper who releases the chute cover himself, uses a rip-cord. He pulls at a handle, not a cord, and it is this handle that releases the covers, which are 'spring-loaded' to whip them away quickly. This loading is provided by stout, elasticated ropes, held under tension until the moment of release.

Toothpaste tubes originally started life as artist's oil colour tubes. They were first patented by artist John Rand (US) on 11 September 1841, and were first used commercially to house the oil colours sold by Devoe and Reynolds. The basic idea was taken up by dentist Washington Sheffield, and in the form of collapsible metal tubes, was used by his Sheffield Tube Corporation (US) as containers for toothpaste (1892). In the same year toothpaste in tubes was first sold in Britain by Beecham's, better known for their pills 'worth a guinea a box'.

Letter boxes were used by a few of the early posts that predated the official British penny post. But the first pillar boxes did not emerge until the idea of adhesive stamps caught on in Belgium. The first Belgian pillar boxes were erected to cater for the expansion that began with the first issue of stamps on 1 July 1849. The same type of box was introduced to Paris in 1850. It was an ornamental cast-iron structure with a protruding lip over each letter-slot to guard against rain. The first pillar boxes in Great Britain were erected in Jersey, in the Channel Isles. The first four in St Helier were very different from the Continental style. They were quite plain but made in a sexagonal form, which gave them an air of distinction. The chosen colour was red and they opened their slots to the public on 23 November 1852. Next year saw the first pillar box on mainland Britain, when the one at Botchergate, Carlisle was put into service. London had to wait until April 1855 before it saw its first pillar boxes, but the capital was rich in post offices, so the need may not have seemed at all urgent at that time.

The credit for introducing these boxes is usually given to Rowland Hill, but the man to first press for them was in fact the novelist Anthony Trollope. At the time he was far from famous and simply a Surveyor's Clerk sent to Jersey to report on the postal arrangements. His report showed that there was no receiving office at St Helier and people had to send mail a mile or more to a main office. Roadside boxes would take care of this problem, as it had in France. This report by Trollope was passed on to the Post-Master General who then authorized the manufacture and erection of the first British boxes. Hill had nothing to do with this activity, but he was very good at taking credits.

The first flag-day was organized at Pontypool and its nearby small satellite Griffithstown on 21 August 1914. Griffithstown in Monmouthshire was a railwaymen's settlement, birthplace of the first raildrivers' union ASLEF. One of the driver's wives tried to please her sons with some type of military insignia for their war games. So she cut up some red, white and blue ribbon and mounted pieces on match sticks, like miniature flags. The local milkman turned up at this point, grew thoughtful and wondered if the flags could be sold to raise funds for the new *National Relief Fund*. This was a charity just set up by the Prince of Wales to aid the dependants of servicemen. The idea met with enthusiasm; sellers paraded their trays of flags through the streets and a new style of charity money-raising began. The woman behind this venture was Mrs Harold George, but no one remembers the name of the wise milkman. Less than two months later saw the first national flag-day, held in aid of the Belgian Relief Fund on 3 October.

Isaac M. Singer, he made the sewing machine commercially possible and launched hire-purchase

Though consumer debts are ancient in origin, the *Hire Purchase System* only came into being with the introduction of Singer's sewing machines. These finely-made machines were priced at $100 so to sell as many as possible fast, seemed out of the question. Many would-be owners simply lacked the ready money. And that is where Singer's partner, Edward

Clark, made history. He drew up a scheme which would allow the buyer to take the machine away without full payment. A deposit would be enough, and the rest could be paid back on easy instalments, tailored to suit the customer. Such a move showed great understanding of consumers' drives. The lure of painless, immediate ownership was too great to resist and HP boosted sales magnificently. This first nationally organized instalment system of September 1856 animated other trades who saw the mass-appeal of such a scheme and emulated. A new, increasingly important factor, then emerged and became part of all economies.

The first tranquillizer taken by man was actually in use at least 3,000 years ago. This was *Snake Root* extract or *rauwolfia* serpentina, first discovered in India and first mentioned in Europe in 1563 by Garcia de Orta. Its value was ignored in Europe, partly because it was wrongly claimed to be a cure-all. This type of exaggerated claim is often found with folk remedies, and unfortunately this can blind people to the true merits of some of the odd medicines found in nature. In the case of Snake Root it was not until 1931 that its properties were scientifically investigated for the first time. The Indian scientists involved, Doctors Siddiqui, Siddiqui and Hussain, isolated five crystalline substances from the root. And a report on its sedative properties was furnished by Doctors Sen and Bose.

By 1949 it was discovered that the root extracts possessed the power of reducing high blood pressure. By 1952, twenty-one alkaloids had been isolated from the root, the most important being reserpine, first isolated by Emile Schlittler and first marketed by Ciba in 1953 as *Serpasil.* Today at least fifty-four preparations used in treating high blood pressure; and in tranquillizing patients with mental and emotional disturbances; are derived from rauwolfia alkaloids, isolated from the Snake Root of antiquity. It is reported that mentally-ill patients treated with reserpine become relaxed, sociable and co-operative, while manic depressives behave as if they did not have a single care in the world.

The mouth-organ grew out of experiments with some of the smaller free-reeds used in the harmonium. The *harmonium*, first developed at the end of the eighteenth century, made its notes by passing air through slots cut in brass plates. Each note had its own slot and fastened above this aperture was a thin, springy tongue of brass, riveted at one end. As air passed through, this tongue vibrated back and forth in its slot and created its special note, which depended on the size and weight of the tongue. Christian Friedrich Buschmann claimed to have hit on this idea of a small mouth-blown organ when making a tuning device based on harmonium reeds. This was in Berlin in 1821, but there is no proof of this claim. We do know that the first mouth-blown reed instruments in Germany operated only when *blown* into. There was no attempt to make anything along the familiar suck-blow principle of the harmonica that we all love or hate.

In Britain a very different type of mouth-organ was invented by Charles Wheatstone in 1829. He called it a *Symphonium* and it took the form of a small lyre-shaped metal box with an ivory mouthpiece and rows of small buttons, one for each note.

From this first British mouth-organ developed the first English *Concertina* in 1849. This was the result of Wheatstone's mature thought. He dropped his mouth-organ as too frivolous and created his *Melophone*, a concertina employing a double action; each button giving the same note whether the bellows was pushed in or out. It was a hexagonal shaped instrument with the notes of the scale given alternately to the two hands. The rival German system allotted the bass notes to the left hand and the trebles to the right hand, and was single action, giving different notes on compression and expansion of the bellows. The name concertina, or rather *Konzertina*, was first used in 1834 in Germany by Carl Freidrich Uhlig of Saxony, to set apart his hand-organs from those made by other makers. The German system is no longer made; all modern concertinas are now made along Wheatstone's system, or the Anglo-German single action system. The Anglo is the system most favoured by folk-singers; the Wheatstone is the first love of the Salvation Army.

The first telephone boxes were not part of the street scene but cubicles set aside in offices. The first of these was constructed on the premises of the *Connecticut Telephone Company* in New Haven, and opened for public use on 1 June 1880. The Call Offices at the Baltic and Stock Exchanges in London (1882) were never open to the public, but were simply there for the benefit of brokers. The first call-offices meant for all were opened by the British General Post Office in August 1884. Payment at that time had to be made to attendants. The first call-box meant to use coin-operation, was placed in the Hartford Bank at Hartford, Connecticut in 1889. This operated under a method patented on 13 August 1889 (No. 408709 US) by William Gray of Hartford. Gray then formed the *Gray Telephone Pay Station Company* in 1891, the first company to rent out coin-box telephones to store keepers.

In Britain the first pay-phone was opened for use in April 1906. This was still not a street box, but an indoor kiosk sited inside the Ludgate Circus Post Office. The first outdoor telephone kiosks arrived in Britain in 1908. These were built by the *National Telephone Company*, but no standard pattern was insisted on. Each one ordered could be made to match in with its surroundings. A photograph of the first erected in Folkestone (1908) shows a rustic gazebo-style structure fashioned from round logs. Only the notice boards, with their huge bell logos clash with the rural background. The first standardized designs were not introduced until 1921 when the GPO erected their first concrete kiosks with red wooden doors.

The first woman to become a Minister of State was Alexandra Kollontai. She came from an aristocratic Russian family but spurned her family and friends and threw in her lot with the Bolsheviks. When the first Bolshevik Government was set up on 27 October 1917 (old calendar), she became the People's Commissar for Social Welfare. She held this post for six months, then resigned because of her total opposition to the way she saw things developing. Her idealism had led her to expect marvels from the revolution, but the hollowness of much of the propaganda she had to listen to sickened her. She then identified with a trend known as the Workers Opposition and her bold, critical views were printed abroad, to the embarrassment of the Soviet Government. In Britain they first appeared in an obscure paper called *Workers Dreadnought* in 1921 (April to August). She was be too well known and too well loved to be executed, so she was virtually exiled from Russia by being sent to the Russian Legation in Norway (1922). She became first woman Head of the Legation in 1923 and in August 1924 became the first woman Ambassador in the world. She spent most of her life in this diplomatic exile until she retired in 1946.

The first heart transplant was performed by surgeon J. D. Hary of Chicago in 1964, but the patient that time was a chimpanzee. The first human heart transplant was carried out by Doctor Christian Barnard at the Groote Schuur Hospital in Cape Town. This operation on 2 December 1967 took six hours to perform and used all the energies of thirty doctors and nurses. But although the patient lived after the exchange, this operation cannot be held to be a success, since the patient died from post-operative pneumonia eighteen days later. The first successful heart transplant was later performed by Barnard in 1968 when his patient, Philip Blaiberg lived on for a full seventy-four days.

The first sighting of a *Flying Saucer* was claimed by Kenneth Arnold, a businessman from Boise, Idaho. He owned a single-engine aircraft and on 24 June 1947 he was flying over the Cascade Mountains of Washington, at a height of 9,200 feet. The weather was fine and sunny, and the trip uneventful until 3 o'clock, when Arnold suddenly saw a blue-white flash. There was a DC-4 on his left but no other craft in sight. Then came a second flash and to the north he spotted a formation of gleaming objects winging over the mountain tops at a great speed. He made an attempt to estimate size, distance and speed and came up with calculations showing that they were twenty miles away flying at a speed of 1,656 miles per hour. On landing at Yakima he told his tale to several of the pilots there, before he took off again for Pendleton, Oregon. When he touched down there, he was met by a bevy of reporters, for this was the time of the Cold War and anything odd or suspicious was newsworthy. When asked to describe the mysterious craft Arnold said 'They flew like a saucer would if you skipped it across the water'.

That one sentence gave the world its first idea of a 'flying saucer'. Within days at least a score of people in different parts of the States made claims to seeing similar saucers, and a new era of strange claims began. So what did he see? One scientist took his calculations and showed that Arnold had judged the objects to be twenty miles away with wingspans of 45 to 50 feet. But the human eye does not have the resolving power to distinguish objects of that size across, at that distance, so the things seen had to have been much closer and were probably military jets at subsonic speed. Their closeness made the speed seem even greater. And there were other valid objections, but nothing stemmed the tide of sightings and there were many more tales to come, many of them quite beyond belief.

The first fire service dates back to Rome of the third century BC. They then had a *Corps of Vigiles* ready and equipped to fight fires. They even had double-cylinder force pumps made of bronze with jointed delivery pipes that could be pointed in all directions. But with the demise of the Roman Empire many of their public services fell into neglect and were even forgotten.

The fire-pump was eventually re-invented, in the seventeenth century. The first fire engine, fitted with a force pump and swivelling delivery tube, is described and illustrated in Heinrich Zeising's *Book of Machines* of 1612. The flexible fire hose invented by Dutchman Van der Heiden, was brought to Britain when William III took to the throne in 1689. This gave firefighters a chance to draw close to a fire, instead of taking long shots from a static cart. At that time there were no municipal firefighting forces. There were small bodies hired by some of the insurance companies who constituted private forces and the first of these was the *Phoenix* fire brigade set up by the first London fire insurance company in 1680. Insured houses carried a plaque on the wall and the firefighters were instructed only to bother with burning buildings if they carried the plaque, or if they were too close to insured premises for safety! Apart from these mercenary minded firemen there were small brigades kept by the City of London Livery Companies. These were set up after the Great Fire of 1666 and were the nearest to public spirited bodies in the Capital until the nineteenth century. Suprisingly enough the first municipal fire brigade in England was organized in the small market-town of Beverley in Yorkshire. This was started in June 1726 and was manned by part-timers who were paid for each call-out.

In the Capital the first full-time brigade arose because of an agreement between ten leading fire insurance companies. The ten decided to amalgamate their private brigades into one large body known as *The London Fire Engine Establishment*. This was first agreed in 1832 and amalgamation first took place on 1 January 1883. First head of this service was Supt James Braidwood. His force was eighty strong. This private enterprise service lasted until 1866 and then it was reorganized as London's first publicly controlled fire-fighting body the *Metropolitan Fire Brigade*. All fire services in Britain were later placed under local authority control. This was replaced in 1941 by the creation of the first National Fire Service, but this was born of wartime needs and was disbanded in 1947, when control passed once more into the hands of counties and county boroughs.

London's first fire brigades were run by insurance companies

The paperback book was first introduced by the Leipzig publishers Christian Bernhard Tauchnitz in 1841. They were English language editions meant for the growing body of tourists who were invading Europe in ever increasing numbers. Tauchnitz's aim was to comply with his contract restrictions which only allowed him to bring out editions for sale and use on the Continent. By making cheap editions he made it possible for people to throw them away when read. And a major cost in publishing at that time lay in having books bound in hard covers. So a paper cover cut costs and encouraged the reader to dispose of it without tears of regret. The first in Tauchnitz's first series, *Collection of British Authors*, was Edward Bulwer-Lytton's novel *Pelham*. All this enterprise had to be limited in scope and we find no trace of any mass-marketing of paperbacks until 1935. That year saw the launch of the first Penguin Books by Sir Allen Lane. First in the series was André Maurois' life of Shelley, *Ariel*, available on 30 July 1935. All these first Penguin paperbacks were priced at sixpence and gave the masses the chance to buy books without having to miss too many meals. After this you can say that the paperback became respectable and revolutionized both publishing and reading habits.

The Texas Oil Baron is one of the clichés of soap operas and cheap fiction, but for many years oil-rich Texas was way down in the league of producers. It was small fry. This changed on Tuesday 1 January 1901; the first day of the first month, of the first year, of the new century. Near a miserable little place called Beaumont stood an oil drilling derrick, the Spindletop rig. For ten years dreamers had sunk their wells in the Spindletop area and no one had got rich. Then in October 1900, other dreamers erected an 84-foot high derrick and starred drilling once more. After six weeks they had a blow-out of gas and water and the drilling gear was damaged. Repairs were made and then they returned back to non-stop drilling but as Christmas drew close they decided to shut down for the holiday and start again on New Year's Day. When they started up again, their drill struck rock and their drill bits simply blunted against it. It was a make or break day. To continue the futile hammering away at rock or just fold up? Like fools they kept on hammering and ten days later the drill pipe suddenly shot through the top of the derrick. A roar of escaping gas deafened the drilling team as they watched mud and slime pour around their engine and boiler. But not a sight of oil; at first that is. Yet there was a promising stirring somewhere below and a sound of subterranean bubbling.

When the oil finally spurted, it plumed up to over 160 feet. With it came rocks and sand that battered at pipes and machinery and left them with no way of controlling the gusher. Controlling the crowds that flocked in was just as problematic; in the end, armed guards had to keep them away from the gusher and its explosive vapours. It took ten days and much heroism before the well-head was capped; and just in time. Within a few days the oil-soaked fields burst into flame, but by then the underground pipes were cemented over and the pipe valve was buried under a mound of earth. The well was saved. The first great Texas oil-field was ripe for exploitation and every Texan began to dream of the fortune lying under his ranch; his farm; his back lot; or even his bed. Many of the dreams came true.

Insulin, that potent weapon against diabetes, was first isolated on 27 July 1921 al the University of Toronto Medical School. The work was carried out by Doctor Frederick Grant Banting and his assistant Charles Best. Within hours of completing the work they made the first tests of the substance on one of the laboratory dogs. In theory insulin should have been able to help a diabetic utilize sugars and lead a fairly normal life; tests on dogs seem to confirm this. But the real test came on 11 January 1922 when a 14-year-old boy was admitted to Toronto General Hospital. His diabetes was in such an advanced stage that there was little hope that the teenager would live for much longer. He became the first human to be treated with insulin, and he survived. Then, helped by daily doses of the miracle cure, he went on to live a normal life. Leonard Thompson was the first of millions to bless Banting and his co-workers.

The Sandwich was not invented by the fourth Earl of Sandwich as most people believe, but by his cook, whose name is lost to history. The Earl was such a devoted and stubborn gambler that he refused to leave the gambling table for lunch. His cook, being a very sensible lady, figured that he would have a much clearer head for gaming if he ate something. So she buttered two slices of bread, imprisoned a thick slice of meat between them, and thrust it into the Earl's hand. He ate it mechanically at first, then with enthusiasm, and so in 1762, his favourite snack was born. It was then christened after its consumer, not its creator. Hardly fair!

The first matches ignited by friction were invented by John Walker, of Stockton-on-Tees in 1826. Walker was a chemist who was looking for easily ignited materials suitable for users of fowling pieces. His first match was accidental; nothing more than the stirring stick he had used to mix up antimony and chlorate of potash. When cleaning the stick, the dried mixture caught light, and Walker realized that he had stumbled upon portable fire-sticks.

His first matches, though, were made of stiff card. Packed in a tin tube, which held one hundred matches, they first went on sale on 7 April 1827. The cardboard was soon replaced by flat wooden pieces, and later on, the tube was discarded in favour of small cardboard boxes. The fame of these fire-sticks spread fast and other chemists began manufacturing their own brands. Since the idea was never patented, they were free to do this. The most famous of his imitators was Samuel Jones of London with his *Lucifers*. Later makers changed over to phosphorous tips.

All these early matches were usually struck on a strip of sandpaper, but they could be fired by striking against walls or any roughish surface. They would even ignite if rubbed against each other and this gave rise to accidents and a concern about the safety of carrying around a box of such sensitive splints. This concern led to the search for a safe match and in 1855 the first safety match was perfected by the Swedish firm of Johan Edvard Lundstrom. In Britain the Lundstrom rights were bought by Bryant and May who placed their first safety matches on sale in August 1855.

The transparent adhesive tape we know as *Scotch Tape* was first prepared in 1925 in the laboratories of the Minnesota Mining Company (the 3M Group). Behind its development lay the need for a masking tape, fit for use in the paint-spraying shops of automobile body makers. Many adhesive papers were tried, but not one of them, when stripped off, was kind to the paint. The researchers finally made a waterproof cellulose tape that met their demanding specification, and this went on sale in January 1928. Transparent sticky tapes reached British shops in 1934, but *Sellotape* itself was not sold in the UK until late 1937.

We think of tracked vehicles, like tanks and agricultural tractors as quite modern innovations; and so they are. But the basic theory of a vehicle-carried track goes back to the eighteenth century. The first patent to set out the requirements for a tracked vehicle was that issued to Richard Lovell Edgeworth on 20 April 1770. This patent (No. 953) reads: 'My Invention consists in making portable railways to wheel carriages, so that several pieces of wood are connected to the carriage, which it moves in regular succession in such a manner that a sufficient length of railing is constantly at rest for the wheels to roll upon, and that when the wheels have nearly approached the extremity of this part of the railway their motion shall lay down a fresh length of rail in front, the weight of which in its descent shall assist in raising such part of the rail as the wheels have already passed over, and thus the pieces of wood which are taken up in the rear are in succession laid in the front, so as to furnish constantly a railway for the wheels to roll upon'. Edgeworth was right in arguing that a system along these lines could overcome the problems of traction on rugged and soft ground; but he never saw his ideas reach fulfilment. He was just too far ahead of his time.

In 1960 a posthumous medal was awarded to the Victorian inventor John Benjamin Dancer. This award, made by the *American Microfilm Association*, recognized Dancer as the true inventor of microphotography. Back in 1839 Dancer had developed a microfilming technique producing tiny detailed photographs that could only be appreciated when examined through a microscope or a high-powered magnifying lens. One of his star demonstration pieces was a series of microphotographs of Queen Victoria and her family mounted beneath the stone of a ring. The stone acted as a magnifier and allowed one to view the minute images. He never patented his idea so other, later workers usually get the credit. The posthumous award was meant to set the record straight.

The first *Teddy Bear* was inspired by a cartoon that ran in the *Washington Evening Star* on 18 November 1902. It was the work of Clifford Berryman and showed President Theodore Roosevelt, gun in hand, refusing to shoot a bear cub. The cartoon was said to be based on a real-life incident during a bear hunt in Mississippi. This hunt was staged during a Presidential visit designed to cool down a border dispute between Mississippi and Louisiana. The caption to the drawing played on this theme, it read 'Drawing the Line in Mississippi'. When the cartoon was reproduced in other newspapers it was spotted by a sweet-shop proprietor in Brooklyn, one Morris Mitchom. Mitchom, a Russian immigrant, combined shop keeping with a small amount of toy making; helped by his wife. They were smitten by the sight of the cuddly little bear and set out to make the drawing come to life. Using brown plush they made a prototype bear, complete with movable arms and legs. This took pride of place in their shop window, and alongside it they placed the cartoon and the legend, *Teddy's Bear*. Family tradition among the Mitchoms has it that Papa Morris then wrote to President 'Teddy' and asked if he objected to the use of his nickname. The President, they relate, wrote back saying, '. . . I don't think my name is worth much to the toy bear cub business, but you are welcome to use it.' Production of *Teddy Bears* began in earnest in 1903, and the cuddly creatures became world coveted.

A rival claim to be the first Teddy inventors comes from the Steiff family in Swabia, Germany. But although they did make similar bears, no proofs of priority have ever been presented by the Steiff Company. And some of their statements are so dubious that they are best forgotten.

The Mitchom claim is both reasonable and acceptable, without straining logic in any way.

Today we take chromium-plated objects for granted, yet it took seventy years of experiments before the first practical plating method was developed. The very first attempts at electrodepositing chromium were made by the German scientist Robert Wilhelm Bunsen, best known to everyone as the inventor of the *Bunsen-Burner*. Those efforts were made in 1854, but it was not until 1924 that the first commercially useful method was perfected. This resulted from the work of Doctor Erix Liebreich in Berlin and was first patented in Britain in 1925 (No. 243,046). When plating steel it was found that a primary plating with nickel would provide an ideal surface layer for the subsequent deposit of chrome. Chrome, by the way, does not rust. The rust spots seen on car bumpers and other metal fittings are always caused by rust from the underlying steel pushing its way to the surface.

Fountain pens of a sort were in use in the seventeenth century. Samuel Pepys used one in August 1663, but none of them were dependable, though convenient, and the quill stayed supreme. The finest of them all was the *Compound Fountain Pen* of Joseph Bramah. This held its ink in a hollow silver tube. The supply to the nib was controlled by squeezing this tube. Bramah's model, first sold in 1809, suffered from a major problem that crippled all fountain pens; the supply of ink was erratic, not smoothly delivered. No amount of redesigning of nibs or ink reservoirs had any effect on this problem. Fortunately, a messy ink-spewing pen put insurance salesman Lewis Waterman in an aggressive mood. This soon shifted over to the thoughtful plane and he began to see that as the ink flowed it needed to be gently replaced by air let into the reservoir. After many disappointments he finally worked out a patentable and reliable answer. An ebonite support under the nib needed to be furnished with grooves or ducts. These would permit ink to flow one way and at the same time allow air to flow in the opposite direction. His patent of 1884 (No. 3,125) gave us the first fountain pen able to write without mishaps.

The first coffee houses were opened in Constantinople in 1554, but these served the thick, syrupy drink now known as Turkish Coffee. A more fluid style of the drink was served in the first coffee house opened in England in 1650. This was at Oxford where a report tells us, 'Jacob, a Jew, opened a Coffey house at the Angel, in the Parish of S. Peter in the East, Oxon, and there it was by some, who delight in

Noveltie, drank'. Two years later the first coffee house in London opened up at St Michael's Alley, Cornhill in the City. The proprietor was a Mr Bowman and the place was called *Pasqua Rosee*. From then on coffee houses multiplied greatly and became especially important in City life. So important that in 1674 vexed housewives organized *The Womens' Petition against Coffee*. This petitioned the King calling on him to close down the coffee houses on two grounds. One, that they were drawing the men away from their havens of domestic bliss; secondly, that when they arrived home late, they were too tired to perform husbandly duties, and in any case were rendered impotent by the noxious drink they were slaves to. All far-fetched, you may think, but King Charles II took them seriously. A Proclamation was prepared announcing the closure of the houses, but popular outcry led to its annulment and this first organized movement to ban the drink lost momentum and expired.

Before the introduction of anaesthetics many surgeons embraced the heartless philosophy that pain was inevitable. French surgeon Alfred Velpeau actually wrote, 'The abolition of pain in surgery is a chimera. It is absurd to go on seeking it today. "Knife" and "pain" are two words that must *forever* be associated in the consciousness of the patient'. These words of 1832 were to be effectively challenged within ten years, by Doctor Crawford Williamson Long, of the village of Jefferson, in Georgia. In the winter of 1841 some travelling showmen had brought a new entertainment to his remote area. Using nitrous oxide or 'laughing gas' they had caused locals to reel around with mirth and dance frantically with wild elation. After the showmen had rolled on, some of Long's friends asked if he could stage a similar show. Long obliged by bringing out a bottle of ether, and when the supplicants inhaled it, they went off into spasms of joyous abandon. So began a series of 'ether frolics' at the doctor's house and at one of these evenings Long bruised himself while in an ether whirl. When sobered up he realized that he had not felt any pain when he suffered the blow leading to the bruises. Using his logic, he then offered to remove two cysts from the neck of James Venables, after administering ether. Venables agreed; the cysts were cut out and the patient felt nothing while the cuts were made.

That first use of ether on 30 March 1842 was followed by a confirming operation on a boy who needed two fingers amputated. One was removed by the traditional methods; the other, after ether had been inhaled. Great pain attended the traditional removal; that performed under ether was mercifully pain free. But Long met with opposition from other doctors, and with superstitious resistance from patients. Threats of a possible lynching led him to forget ether for good. So brute ignorance

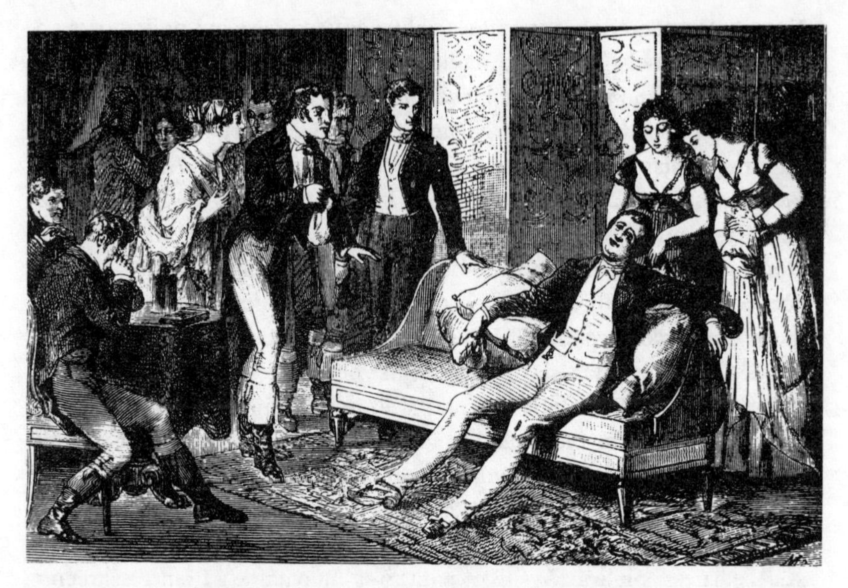

First came 'laughing gas' parties, then 'ether parties'.
These led minds to see the medical possibilities

triumphed, and it was left for William Morton to make the real breakthrough.

Morton was a dentist who had experimented with ether, and had met with uneven results. Sometimes the fluid worked as planned, but at other times it gave unwanted problems. Puzzling this out Morton realized that the poor results had been caused by unrefined ether. The impurities mixed in with the low-grade ether were the cause of failures. Only a pure, rectified ether would work miracles. On 30 September 1846 Morton shut himself in his room, sat in his dental chair and saturated his handkerchief with rectified ether. Seven to eight minutes later he came out of his unconscious state and realized that he could have had a tooth drawn in that time without knowing a thing about it.

Later that night, long after surgery hours, his doorbell rang and an agonized man stood there asking for help. His toothache was unbearable and he begged for an immediate extraction. He led the man to his chair, told him about the wonders of ether and the man agreed at once to having it administered. The tooth presented no problems. It was extracted swiftly; the unconscious man was brought around fast with a cold water shock to the face, and the first dental extraction became a matter of record. This first patient, Eben Frost even wrote out his own account, saying: 'This is to certify that I applied to Dr Morton at nine o'clock this evening (September 30, 1846), suffering under the most violent toothache; that

Doctor Morton took out his pocket handkerchief, saturated it with a preparation of his, from which I breathed for about half a minute, and then was lost in sleep. In an instant more I awoke and saw my tooth lying on the floor. I did not experience the slightest pain whatever. I remained twenty minutes in his office afterward, and felt no unpleasant effects from the operation.'

The first public account of Morton's work was published in the Boston *Daily Journal* for 1 October 1846. Within days of his triumph Morton began canvassing the support of surgeons. His method could alter their techniques, take away their anguish and bring hope to their patients. But he met resistance. He was a mere chemist. His method most likely had hidden dangers. Their skills had worked for centuries. One man listened though. Doctor Henry Bigelow of the Massachusetts General Hospital liked what he heard and introduced Morton to the most eminent surgeon in New England, Doctor John Collins Warren. Warren wasted no time; on 16 October 1846 he summoned Morton to stand by him while he operated. The patient was a young printer, Gilbert Abbott, with a tumour under his jaw. Morton gave the patient ether, using a special inhaler he had designed. The patient sank into a state of oblivion and Doctor Warren began the operation. The audience of doctors were awe-struck by the silence. No screams, no threshing of limbs. The patient lay there in the deepest sleep. In thirty minutes the first major operation held under a state of anaesthesia was over. And the patient was well enough to leave hospital on 7 December.

There were still many things to be learned before full control of Morton's method was realized, and a good name was needed; people just could not go on referring to it as 'gas' or 'the mixture'. Morton himself favoured 'letheon', a reference to the river Lethe of Greek mythology; one drink from this river would wipe out all painful memories. Oliver Wendell Holmes, poet, and professor of anatomy at Harvard, disagreed. On 21 November 1846 Holmes first gave the world the definitive name for Morton's method. Holmes wrote, 'Everyone wants to have a hand in a great discovery. All I do is to give a hint or two, as to names . . . The state should I think be called "*Anaesthesia*" . . . The adjective will be "*Anaesthetic*" . . .' And so it was.

But the newly named blessing soon attracted acrimony when Doctor Charles Jackson suddenly claimed to have been the first to discover the value of ether. He declared that he had made the discovery five years before Morton. Why then had he let patients suffer for those five years? To this question Jackson had no answer; but that did not stop him from hounding Morton for years. British surgeons first learned of American anaesthesia in November 1846, when a letter from Doctor Henry Bigelow's father to

Doctor Francis Boot brought them the news. Then on Saturday, 16 December Doctor Boot gave ether to his niece while dentist James Robinson extracted a tooth. Fully satisfied with the results; he then sent Bigelow's letter on to surgeon Robert Liston, fastest man with the knife in England. Liston relied on speed to cut down shock; but he was not a bigot and he set up an amputation, due on 21 December, as a test of the anaesthetic method. On that day, at University Hospital London, Liston performed the first major operation under ether. In twenty-seven seconds he amputated a diseased leg and the patient lay there, still flat out and with no visible signs of distress. No one watching that day had ever seen such an operation. Their memories were of screams, terror and suffering. The first man to benefit from anaesthesia in Britain was butler Frederick Churchill.

That night Liston staged the first celebration of the arrival of anaesthesia with a dinner party at his house at 5 Clifford Street, off Old Bond Street, London. First details of this conquest of pain were carried in the *People's London Journal* on 9 January 1847. Soon the whole of Europe knew of the new hope. There was resistance; yet in France, even Velpeau was persuaded to use ether; and he took back those chilling words of his uttered years before. Pain was there to be conquered; not endured as a penalty imposed by a deity seething with wrath – an argument seriously advanced by some warped creatures.

Microbes have been used for centuries in food production. Cheeses, yoghurts and buttermilks depend on them; wines and ales would be lost without them, but in all these cases the microbes are used as auxiliaries. Only brewed yeasts were used as foods in their own right, whether in soups, sausages or drinks. Now, that has changed. For the first time microbes are being cultivated as a food stuff proper in the shape of SCP products. SCP (single-cell protein) production first started off in the 1960s when animal foodstuffs were derived from microbes grown on oil-based chemicals. First of these foods was *Toprina* made by British Petroleum at their plant in Sardinia. ICI then followed with their *Pruteen* feed made from the bacterium Methylophilus Methylotrophus.

It was left to Rank, Hovis, McDougall to go one step beyond and make the first mycroprotein meant for human consumption. They used a species of the Fusarium mould which grows easily on carbohydrates, such as potatoes, wheat, rice, cassava, and so on. *Fusarium* has a ratio of about 45 per cent protein to 13 per cent fat, and that makes it as nutritious as many meats. It was first tried out on hundreds of people both within the manufacturing company and without. There were no unexpected side-effects, nothing but approval. After that, it was fed to animals and after extensive tests was shown to have no harmful effects.

In 1986 the new food was placed on sale as *Quorn* and used in pies, pasties and other handy forms. This first cultivated micro-protein was originally meant to substitute for meats and other expensive proteins, but as of today, the price of *Quorn* is still too high to make it attractive to anyone except dedicated vegetarians and vegans, or food dilettantes.

The first electric razor was invented by Colonel Jacob Shick (US) in 1929 and patented by him in 1930 (No. 326,374). Production of the razor began in 1931 but there was no overnight rush of customers. Many doubted that it could give them the close shave and clean feeling produced by wet shaving. Others found the price far too high. To own one, would cost more than a few years supply of blades. Converts came in small batches at first, then they acted as missionaries and spread the good news. By the end of the 1930s Shick's sales were counted in millions each year.

As you know, the *Statue of Liberty* has a viewing platform placed around the flaming torch held aloft by the right hand. The first people to climb to that platform and look down at America did so in June 1876. But the view they saw was rather different from the one now on offer. Beneath them they saw the metal fingers, the metal wrist and a little of the forearm. Then came the entrance hall through which they passed! At all sides were the grounds and pavilions of the Philadelphia Centennial Exposition, for the French makers of the statue were behind schedule and the hand was the only part ready in time for the celebrations.

As the French craftsmen toiled on under the guidance of the designer of the statue, Frédéric Bartholdi, an odd attitude developed in the USA. We now think of the statue as the symbol of the great American Dream, the light of hope for the 'huddled masses yearning to be free'; but in 1885, Congress was unwilling to appropriate money to provide the statue with its pedestal. It was left to the New York *World* to raise the $100,000 needed, through an appeal to its readers. On 26 October 1886 the bronze *Statue of Liberty* was first unveiled at Bedloe's Island, New York Harbour. The statue was a free gift from the French Government. Its pedestal a gift from members of the public ashamed of the parsimony of their political masters.

The first authentic perambulator was ordered by the Third Duke of Devonshire from the carriage maker William Kent of Chatsworth. It was completed in 1733 in the form of a small, four-wheeled coach. This carriage still exists and the draw-shafts and collar found on it show that it was originally meant to be drawn by a dog, not pushed in modern

The Statue of Liberty was sent from France in pieces.
The hand and torch arrived first

fashion. The first push-chair perambulator was not made until 1850 when two competing firms, John Allen and A. Babin, both of London, produced very similar models using three wheels instead of four.

The first electric organ was developed in 1930 by French inventors Messrs. Couplex and Givelet. It was a pipeless organ in which 'Pipes are replaced by radiophone lamps, which give all the timbres of the various stops. For varying the intensity and qualities of sound this new organ is fitted with amplifiers'. Despite the enthusiasm of the inventors, the eighty oscillators used in their organs made them hard to keep in good order. Because of this, the first electric organ to become commercially successful was the *Hammond organ*, invented by Chicago clockmaker Laurens Hammond, and introduced in 1935. In one month, December 1935, fifty-one American churches installed Hammonds.

The first Ether-Wave musical instruments, as distinct from electric organs, were devised by Russian scientist Professor Leon Thérémin in 1927. First called the *Etherophone*, but better known as the *Thérémin*, it was a large rectangular wooden box sprouting two metal aerials. One was a simple perpendicular rod; the other a horizontal loop. Each aerial was linked with its own electric circuit and bank of radio valves. The valves were placed in a state of oscillation and the pitch of this oscillation was controlled by moving the right hand towards or away from the rod. Volume was controlled by passing the left hand over the area above the loop.

The inventor demonstrated his instrument in many of the world's capitals and won over many serious musicians. They were enthralled by this new, eerie musical voice and in 1929 the first suite was written for the *Thérémin* by composer Joseph Schillinger. This was his *First Airphonic Suite*. In the same year the first Thérémins were marketed by the great record company, RCA Victor of Camden, New Jersey. They also sold the first combined Thérémin and electric gramophone in 1929, and added an extra dimension to owning one, by issuing two sets of special records. One set gave instructions for playing the instrument and provided scale and arpeggio passages which could be imitated. The other set gave a series of accompaniments that could be played while the Thérémin provided the melodies. This was the first musical tuition course conducted by records.

A rival ether-wave instrument first appeared in France in 1928. This was the *Ondium Martenot* invented by musician Maurice Martenot of Paris. It differed greatly from the Thérémin and in its most used form, employed a dummy keyboard to control pitch. The left hand was able to control the tone-colour of the notes by switching various filters in and out of circuit. On its first public appearance in 1928 the first piece written for it was performed. This was Dimitri Levidis' *Poème Symphonique pour Solo d'Ondes Musicales et Orchestra*. Soloist at this time was the inventor, who then took the instrument to America and gave the first US performance of the *Poème* with the Philadelphia Orchestra, conducted by Stokowski (1930). After that some thirty composers wrote for the instrument. Best known of these was Arthur Honegger who used it in his *Semiramis* and his *Joan of Arc at the Stake*.

In 1818 the British economy faced a crisis brought about by the wear and tear on the plates used for printing banknotes. Increasing numbers of notes were in circulation but security printing was hardly possible since the engraved copper plates were so soft that only short runs were possible. Then the plates had to be re-engraved, and that introduced variations in the patterns. As a consequence there were no absolute standards to match notes against and that was a boon to every eager forger.

In 1818, the American Jacob Perkins, a prolific inventor and a superb craftsman, came to Britain, settled and introduced the first security printing system based on the use of hardened steel engraved plates. By engraving on soft steel, then tempering it, he proved that he could take up to 30,000 prints from each plate. To make this method even more economical, he then invented the first transfer press. The design for the note was divided up into sections. One engraver would engrave a value, another a letter, yet another a vignette and so on. Each engraved section was then placed in position on the bed of the transfer press. A soft steel roller was then passed over the engravings and the recessed lines were picked up as raised lines. When hardened this roller could then be used to create even more printing plates. The excellence of this new printing method made his firm of Perkins, Bacon and Company, first choice when the British Government was seeking a printer for an entirely new product; an adhesive postage stamp. The Penny Black first sold on 6 May 1840 was a Perkins product. His transfer system lay behind the security achieved by the complex, yet beautiful, design.

But it was not all ink and print with Jacob. In between banknotes and stamps he took time off to invent an extraordinary weapon; the world's first steam gun. He demonstrated this on 6 December 1825 at Regent's Park, London, before the Duke of Wellington and a batch of high ranking army officers. His gun was powered by a steam boiler working at 900 pounds per square inch and it was capable of discharging 1,000 bullets a minute! When tested its bullets penetrated no less than eleven one-inch deal planks. The Duke of Sussex thought it 'Wonderful, damned wonderful', but the army advisers held that the steam boiler would make it unwieldy on a battlefield and they told Perkins to steam off. So he did, and bested them all by keeping cool and registering a patent for the first practical refrigerator (No. 6,662 1834).

• INDEX •